THE HUMANS WHO
WENT EXTINCT

THE HUMANS WHO WENT EXTINCT

WHY NEANDERTHALS DIED OUT AND WE SURVIVED

Clive Finlayson

OXFORD
UNIVERSITY PRESS

OXFORD

UNIVERSITY PRESS

Great Clarendon Street, Oxford OX2 6DP

Oxford University Press is a department of the University of Oxford.
It furthers the University's objective of excellence in research, scholarship,
and education by publishing worldwide in

Oxford New York

Auckland Cape Town Dar es Salaam Hong Kong Karachi
Kuala Lumpur Madrid Melbourne Mexico City Nairobi
New Delhi Shanghai Taipei Toronto

With offices in

Argentina Austria Brazil Chile Czech Republic France Greece
Guatemala Hungary Italy Japan Poland Portugal Singapore
South Korea Switzerland Thailand Turkey Ukraine Vietnam

Oxford is a registered trade mark of Oxford University Press
in the UK and in certain other countries

Published in the United States
by Oxford University Press Inc., New York

British Library Cataloguing in Publication Data

Data available

Library of Congress Cataloging-in-Publication Data

Finlayson, Clive, 1955–
The humans who went extinct : why neanderthals died out and we survived / Clive Finlayson.
p. cm.
Includes bibliographical references.
ISBN 978–0–19–923918–4 (hardback)
1. Neanderthals. 2. Human evolution. 3. Social evolution. I. Title.
GN285.F54 2009
303.4—dc22   2009020070

Typeset by SPI Publisher Services, Pondicherry, India
Printed in Great Britain
on acid-free paper by
CPI Antony Rowe, Chippenham, Wiltshire

ISBN 978–0–19–923918–4

1 3 5 7 9 10 8 6 4 2

Contents

List of Illustrations

Preface

WHY did the Neanderthals go extinct? I often get asked this
question when I give a public lecture. The answer most people
expect is that our own kind, those referred to as Modern Humans by
palaeoanthropologists and simply as Ancestors, or *Homo sapiens*, in this
book, wiped them out; it could have been overt aggression or a more
subtle competition but eliminate them we did. Behind these assertions
is the perception that 'our people' were cleverer than 'the others', those
ape-like Neanderthals. So it is obvious that when we both met there
could only have been one outcome. We are here after all and they
are not.

I started to question this dogma almost a decade ago now and I
have become increasingly dissatisfied with this simplistic explanation.
I started by chasing the evidence and found that there was none. When
I challenged the defenders of this view at conferences the best answer
that I got was that in all archaeological sites that had been studied
H. sapiens remains and artefacts always appeared above, and therefore
later than, Neanderthal (*Homo neanderthalensis*) ones so it was clear
that *H. sapiens* had come in and kicked *H. neanderthalensis* out. When
I suggested that the same evidence could be interpreted to mean that
H. sapiens were only able to enter these caves once the Neanderthals
had gone, and by implication that the Neanderthals had actually kept
them out, there was silence.

The point is that sometimes one piece of evidence can be interpreted
in more than one way, in the process raising doubts about uncorrob-
orated assertions. The staunch crusaders of the idea that *H. sapiens*

actively replaced *H. neanderthalensis* continue to cling onto this idea in the face of increasing evidence against it. But that is part of the process that is science.

Who were the Neanderthals? They were humans who separated from our lineage around half-a-million years or so ago (the precise time is uncertain). In this book I will treat the two, for convenience, as separate species (Neanderthals, *H. neanderthalensis* and Ancestors, *H. sapiens*) since they reperesent two distinct lineages that appear to have been geographically isolated—*H. neanderthalensis* in Eurasia and *H. sapiens* in Africa—for a considerable time. But this should not be taken to mean that I am convinced that the degree of distinctiveness between the two merits giving them such a status. Other authors consider them subspecies of *H. sapiens* (*H. sapiens neanderthalensis* and *H. sapiens sapiens*) but the acid test—did they interbreed and therefore behave as a single biological species?—is hard to show from fossils.

Since the separation of the two lineages, our people and the Neanderthals developed differences that probably had to do with their lifestyles. The Neanderthals became a tough, well-built, people. They had large brains, even bigger than ours, and they lived across Europe and northern Asia as far as eastern Siberia and perhaps even into Mongolia and China. They could probably speak and they were highly adaptable; in some places they ambushed deer and even larger animals while in others they beachcombed or gathered pine nuts. Rarely would they have taken on the largest animals—the image of the Neanderthal taking on a woolly mammoth is probably false. They probably scavenged these giants instead, chasing off wolves and hyaenas in the process. The Neanderthal lifestyle worked for tens of thousand of years.

Our ancestors came from Africa but the routes that they followed, now much clearer since genetic markers have been used to trace the pathways, were not straightforward; we will look at these in this book. There are enigmas waiting to be resolved along the way: why did these people reach Australia almost 15 thousand years before they got into Europe which is much closer to Africa? Were the Neanderthals keeping them out? It is now quite clear that the big entry of *H. sapiens* into

Europe and across Siberia started off in Central Asia: most Europeans, Native Americans, and eastern Asians come from this stock.

There are also tantalizing glimpses that suggest increasingly that other people may also have been around. Why should it only have been Neanderthals and our ancestors? As we begin to understand the complex panorama of prehistoric humans in greater detail we may be surprised to find that the diversity of peoples, from populations to species, was much greater than the simple *H. sapiens–neanderthalensis* dichotomy which we have inherited. The discovery of the Hobbits, *H. floresiensis*, on Flores is the tip of the iceberg.

But we still have to answer the question 'Why are we here and not the Neanderthals?' I am afraid that my answer is not as simple as 'we clubbed them on the head.' The answer is actually a series of answers and, even though we are much closer today than we have ever been to resolving the question, these answers are incomplete. At one level the drastic climate changes that hit the parts of the world where the Neanderthals lived after 70 thousand years ago decimated and fragmented their world.

Because the stocky body of the Neanderthal had been interpreted as an adaptation to cold climate, the idea that the cold had a negative impact on them was not given serious consideration. But body proportions are not only linked to climate: in the case of the Neanderthals it had more to do with their hunting style and, in any case, when conditions were really cold in northern Eurasia the Neanderthals were simply not there anyway. For over 40 thousand years constant attrition by cold environments took its toll and it is to their credit that the Neanderthals held on for so long. In comparison, our people never had it so rough for such a length of time. The last Neanderthal populations, scattered in southern Iberia, Crimea, the Caucasus, and other remote haunts, were like endangered populations of giant pandas or tigers of today. They slowly vanished, one by one. By then they had become 'living dead' and the stories of the disappearance of each nucleus were probably very different: disease, inbreeding, competition, random fluctuations in their numbers.

Huge progress has been made in recent years, particularly in the field of genetics and the study of ancient DNA. We know now that Neanderthals were pale skinned, had a range of hair colours comparable to Caucasians, and had a gene that we share and which is involved in language. More revelations will follow as we move towards the publication of the Neanderthal genome and we may even finally resolve the question of how frequently Neanderthals and our ancestors had sex with each other.

This book is also about our own people. Why did we make it? My answer is a combination of ability and luck. Sure, we were good at what we did but we were also very fortunate to have been in the right places at the right times. Of course, there would have been other people of the same stock who were just as good but ended up in the wrong places at the wrong times and, like the Neanderthals, also went extinct. And that, for me, is a sobering thought that puts me in my rightful place in the cosmos!

Many people have helped me along this voyage of discovery. It all started in 1989 when I met Chris Stringer and Andy Currant of the Natural History Museum in London when they were getting interested in Gibraltar's caves, but there was already by then a parallel story. It was one that started me off in life, as a student of nature: I have spent most of my life studying birds and their ecology. I still do. Many of the insights into the way of life of Neanderthals and our ancestors have come precisely from an understanding of ecology and how the natural world works. My ideas, that I have tried to faithfully put forward in this book, are the product of ecology, archaeology, and anthropology.

My wife, Geraldine, has been my partner in this adventure and to her I owe much of the discussion; I thank her for keeping my feet on the ground and for preventing me from straying too much. And my son, Stewart, has been an inseparable companion in the field, just as my own father was when I was starting off.

A number of friends have commented on parts or the whole manuscript which has improved as a result: Darren Fa, Pepe Carrión, Marcia Ponce de León, and Christoph Zollikofer. The greatest blessing that getting into this field has given me is the friendship of some wonderful colleagues: to those that I have already mentioned I proudly add Kimberly Brown, Paco Giles Pacheco, Joaquín Rodríguez Vidal, Larry Sawchuk, Mario Mosquera, Esperanza Mata Almonte, Paqui Piñatel Vera, José María Gutierrez López, and Antonio Santiago Pérez.

Last, but by no means least, I am most grateful to my editor at OUP, Latha Menon, for supporting the germ of the idea that was to become this book and for helping me shape it.

To all of them I dedicate this book.

When Climate Changed the Course of History

C ONTRARY to popular belief, history does not repeat itself. The story of our planet was not predetermined, there was no air of inevitability to it, and the story of life does not speak to us of a linear progression from primitive to sophisticate. Instead, its shape has been carved out by the accumulation and loss of information, genetic and cultural, creating the illusion of relentless progress. This history is full of cases in which chance events radically altered the world and changed its course. Had these random proceedings not happened in the time and place in which they did, I would certainly not be writing these lines today nor would you be reading them. This book is about one of countless stories of life on Earth; it is of particular concern to us, not because it was in any way extraordinary but because it involved our own species.

The world is full of successful living beings some of which have persisted, barely changed, for millions of years. They have been the lucky ones because the way that they adapted to their present inadvertently made them successful in their futures. We might arrogantly think of ourselves as members of the exclusive survivors' club but, in reality, we are mere novices in comparison to some of the tough designs out there. Even they are the exception though, most species having gone out of business at one time or another during the course of the Earth's haphazard history. As circumstances have changed— continents moved, mountains grew, seas receded, ice caps expanded, climate dithered—most species went extinct and new ones captured a slice of the market. Among the new ones were several kinds of human. The Neanderthals were one of them and they became a highly successful people who managed to live in the increasingly inhospitable world of Europe and Asia for over 300 thousand years, a lot longer than the period that covers our own time on this planet. One day the Neanderthals shared the fate of millions of other life forms and died out. This book is the story of the Neanderthals, how they came to be so successful and how they eventually vanished. How was it that an intelligent and triumphant kind of human could become so vulnerable to external forces that it went extinct? But this is also the story of our own kind, a parallel dynasty of humans that shared parts of the planet with the Neanderthals for a while. I will explore what happened when Neanderthals met our Ancestors and I will attempt to answer the burning questions: did they interbreed? Were Neanderthals really dumb brutes incapable of behaviour that we would consider modern? Did our Ancestors finish them off or did climate change have anything to do with it? But this journey will take us much further. I hope that, by contrasting the Neanderthals to us, we may be able to carry out an auto-inspection. Ultimately, this is also a vision of how and why we are here today and the Neanderthals are gone.

Climate is a key ingredient in the story—it was the architect that moulded our intelligence, our biological makeup, in fact everything that made us human, but it was also the cause of hardship and extinction. Serendipity is central to the argument underlying our account: people, who were in the right place at the right moment, even though at the time they did not know it, got lucky. Others were not so fortunate and cannot be here to share the story with us today. It might so easily have gone another way: a slight change of fortunes and the descendants of the Neanderthals would today be debating the demise of those other people that lived long ago. This is not a trivial question. Behind it lies the implication that we are not as unique and special as we might think. We owe our existence to a series of events in which chance played a huge role. It is sobering to think that there have been alternative ways of being human, that some of the options vanished despite good design, and that such a fate might have easily awaited us round some unexpected corner of our short history. Indeed, it may await us still.

Before we plunge into the question of the Neanderthals and the Ancestors,[1] we need to pause and situate ourselves. By observing the deeper timescape we can grasp the sequence of circumstances that would lead to the meeting of two human populations one distant day in the icy lands of Pleistocene Europe. This prelude was a long one, taking up many millions of years, but it is one that we cannot ignore as it provides the context for later events. It will take up much of this chapter and the next two, during which I hope to capture the vastness of the timescales involved in our evolution. We will encounter, during this long journey, many of the key factors that later on had an impact on the lives of Neanderthals and our Ancestors. We could justifiably select many different starting points to the story. A deep past would take us back to the very origins of life, while a less remote moment would be the actual origin of our most direct ancestors around 200 thousand years ago.

Both choices would be appropriate, as would a number of other intermediate landmarks, but for me the natural starting point is a cataclysmic event that shook the Earth 65 million years ago with far-reaching and irreversible consequences. A massive asteroid impact together with major volcanic activity and sea level changes caused the extinction of all land animals larger than a small dog (the K/T event). It included the dinosaurs and it opened a window of opportunity for other animals. Our early mammalian ancestors were among those that seized the moment, unwittingly paving the way for the future emergence of the primates; but how did we get to an intelligent primate from a tiny, shrew-like mammal that spent its life scurrying in the undergrowth of some remote and ancient forest?

For a long time, the diversification of the mammals from these early ancestors to the spectrum of shapes and sizes familiar to us was thought to have taken place once the dinosaurs had disappeared. Their extinction opened up opportunities that permitted mammals to take on new jobs. But the story unfolding is more complex than we had originally imagined. In recent years fossils discovered in China, Madagascar, and Portugal are showing that mammals had already diversified beyond small, unspecialized, omnivorous creatures long before the K/T event.[2] Aquatic mammals and medium-sized carnivores (one individual fossil was found with a small dinosaur in its stomach!) were already around between 170 and 120 million years ago. These may simply have been early experiments that also met a sudden end at the K/T boundary but the door of the chapter of the early diversification of the mammals must remain ajar awaiting more fossils. Right now we can only speculate on what kinds of mammals might have emerged had these early prototypes made it through the K/T gauntlet. Our story might well have been very different, or it may never have happened.

But it did happen, and after 65 million years ago mammals, which we would find familiar, began to appear on the scene (see timeline in Figure 1). These included early primates around 60 million years ago. They were small, squirrel-sized animals that owed their success to another major global climate change. The boundary between the Palaeocene[3] and the Eocene around 55 million years ago was marked by 100 thousand years of powerful global warming, on a scale that would not be repeated. This rapid warming, raising sea surface temperatures by up to 8°C in a mere 10 thousand years, permitted evergreen forests to stretch across high latitudes of the northern hemisphere, and these provided ideal habitat for the early tree-dwelling primates. A recent study of these fossil primates has given us a detailed picture of how this geographical spread took place:[4] starting in southern Asia, they spread north-east into North America and, from there, round into Europe using land connections long gone. This was the first global expansion of primates and climate was the catalyst.

Another 25 million years would pass before anything remotely resembling an ape would make an appearance. In the intervening years the world would change dramatically from greenhouse to fridge. Winter frosts made their first appearance in high latitudes where subtropical forests, alligators, and flying lemurs had once thrived. An ice sheet

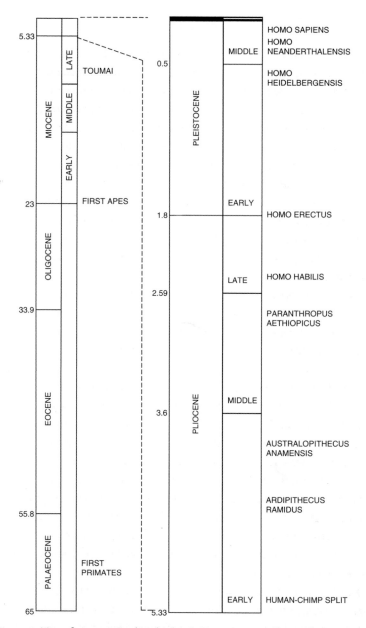

Figure 1 Time frame covered in this book. Dates are in millions of years. Some key events and species are indicated as reference markers. The last 5.33 million years are expanded in the lower column. In this timeline, and throughout this book, the conventional Plio-Pleistocene boundary at 1.8 mya is adopted. Some recent workers have reset this boundary at 2.6 mya

replaced temperate rain forests in Antarctica around 36 million years ago, coinciding with a sharp drop in global temperatures, and tropical forests became restricted to low latitudes. In North America the mean annual temperature dropped by 12°C and many species went extinct.

To find the causes of this radical turn of events we need to look at the movement of the world's major land masses. Gradually the continental plates had been drifting into their present positions but a number of dramatic events were about to unfold. Most significantly, around 54 million years ago India crashed into Asia. This impact generated the lifting of the Himalayas and the Tibetan Plateau five kilometres into the sky, a process that continued until at least 15 million years ago. The consequences were global. The Tibetan Plateau, approximately half of the United States of America in area, began to influence atmospheric circulation patterns. The jet streams were deflected, monsoonal circulation intensified and rain fell heavily on the slopes of the Himalayas. One consequence of the exposure of young rocks to the atmosphere, as the land was lifted and in combination with the increased rainfall, was a high level of chemical weathering. Carbon dioxide was removed from the atmosphere as it dissolved with the rocks. Lowered carbon dioxide levels in the atmosphere in turn caused global temperatures to drop.

Major tectonic events, operating at timescales of millions of years, were largely responsible for the large-scale climate changes that unsettled the planet. To the uplift of Tibet we can add massive volcanic activity in the seabed of the North Atlantic; the opening of two Antarctic marine gateways—the Drake Passage between Antarctica and South America and the Tasmanian Passage between Antarctica and Australia—as these continents separated; the formation of the Andes and the Rockies; and the closing of the Central American Seaway in Panama. These changes were irreversible—once Tibet was elevated, for example, there was no going back—and so, at these large timescales, climate change tended to go in one direction, in this case taking the form of a long-term cooling trend. At shorter timescales (tens to hundreds of thousands of years) regular changes in the Earth's orbit, its tilt and its rotation, meant that the amount and the timing of solar energy reaching different parts of the planet varied, causing repetitive and alternating climate cycles. These accounted for cyclic periods of warming and cooling which we shall see in sharper focus when we look at the glaciations of the past two million years.

It is now emerging that, in addition to the changes that I have described, rare and very rapid shifts of climate extremes (at scales measured in thousands of years),[5] have had profound long-term effects on life. The sharp global warming at 55 million years ago, that triggered the expansion of the early forest-dwelling primates, was the most striking but two other perturbations, at 34 and 23 million years ago, involved cooling. The first of these events, a 400-thousand-year glacial, coincided with the appearance of large continental ice sheets on Antarctica and involved major changes in ocean circulation. The second event was shorter lived (around 200 thousand years) but it was very intense. These climatic anomalies reveal the unpredictable nature of climate changes as orbital, atmospheric, and tectonic factors came together, often in unexpected ways.

So the long period between the global warming of 55 million years ago and the appearance of the first ape-like creatures around 23 million years ago was one of steady climatic cooling; it was closely linked to the final rearrangement of the continents as random terrestrial events conspired with cyclical astronomical rhythms. At the end of this long period the position of the land masses had taken a familiar look. In the process the polar ice caps had made an appearance and had grown; two short and intense glacial events had transformed ecosystems and animal communities; sea levels had fallen conspicuously; polar broadleaf forests disappeared and tropical forests shrank; herbivorous mammals became common. Amidst this turmoil primates went into decline and were restricted to areas of suitable habitat near the equator. The early primates that had been so successful during the earlier period of global warming lost out as their forest habitats retreated. Those that held out would move on to the next episode of this unpredictable story.

The ancient hothouse world got a temporary reprieve between 23 and 15 million years ago when climate returned to its former warmth.

It would not last though, and the downward trend in climate continued inexorably towards the present. For a brief moment life in the early part of the Miocene recalled the planet's former glory. The climate was warm and humid, with tropical and subtropical forests expanding within Africa, and even into former haunts across Eurasia as far as eastern Siberia and Kamchatka.[6] It was the chance the primates had been unconsciously waiting for, but this time they were very different in appearance from those of the earlier period of global warming. These animals were by now distinctively apes.

The opportunities presented by the new African forests meant that apes were now able to eat a wider range of foods than before.[7] Not all predominantly ate ripe fruits, the main diet of most modern apes; leaves, nuts, and other plant material became a regular feature of the diet of some of these apes. They were all similar to each other in one way: they shared a body plan, one that confined them to moving along tree branches on all fours, but their joints were extremely flexible, something unheard of until that point.[8] That flexibility in the joints would, in time, allow apes to adopt diverse forms of movement, such as hanging and swinging from branches using their forearms. Crucially to our story, it would give one distant descendant the ability to manipulate the hand and, among many other things, make tools. For now, neither the ape nor the conditions were right and humans were as far away in the future as the early primates were already in the past.

Today we must travel to remote forest haunts in Africa and southeast Asia to find apes. The situation was similar between 23 and 17 million years ago except that the apes were restricted just to Africa. There were no apes in Asia. Africa, where these apes had evolved, was not joined to Eurasia at that time and the apes could simply not get across.

Around 19 million years ago the African and Arabian plates collided into Eurasia, closing the seaway that had kept them apart. For a period of around 5 million years the land bridge repeatedly opened and closed as sea levels rose and fell. After that Africa, Arabia, and Eurasia were linked by land right up to the present. Today, we distinguish Africa,

Europe, and Asia as separate continents but the reality is that the huge land mass has been a supercontinent for the past 14 million years. This artificial division of Africa from Eurasia has conditioned our view of human evolution as we will see throughout this book and it is a distinction that I am keen to erase.

Even though the land connection established 19 million years ago allowed animals the size of elephants to get across into Eurasia (others also went the other way into Africa) the first apes did not venture north until around 16.5 million years ago.[9] One hint as to why it took so long for apes to exit Africa comes from the teeth of these early immigrants: the teeth had a thick enamel covering which would have allowed them to process hard foods such as nuts. This innovation seems to have allowed them to become independent from fruit and to foray into a wider range of woodland environments and geographical regions. Once again it was a question of having the right makeup as well as the access route to get across. We should not, however, see this as an early form of 'out-of-Africa'. It was, instead, a geographical expansion into suitable environments, some of which happened to be on the African, and others on the Eurasian, part of the land mass.

These apes thrived on reaching unoccupied territory and they spread far and wide across Eurasia. At times they were separated from their African relatives as the land bridge was swallowed by the rising sea; however, by then they had gained an important foothold on the other side and continued to flourish. It is very often the case that animals reaching new and unoccupied lands are able to rapidly differentiate into several types that are able to exploit different segments of the virgin environment. The basic body plan of the colonizers then gets modified into a range of prototypes. The phenomenon is often seen on islands, such as the Galapagos, where Darwin observed one such case of adaptive radiation in the absence of competitors from a common finch plan. We observe a similar adaptive radiation among the ape colonists of the mid-Miocene world.

Orang-utans, gibbons, chimpanzees, gorillas, and humans are the lone survivors of the mid-Miocene ape apogee, around 16 million years

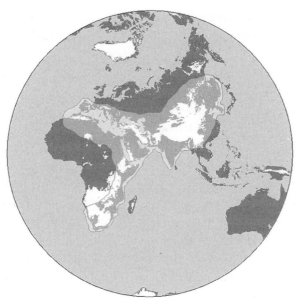

Figure 2 Approximate distribution of wooded environments suitable for Miocene apes between 14 and 9 million years ago

ago, when apes of many different shapes and sizes lived from the Iberian Peninsula to China and from Kenya south to Namibia.[10] This was a vast region of tropical and subtropical woodlands, difficult for us to imagine today, and the apes spread out right across it. In some cases they went from Europe into Africa and in others in the opposite direction. The distinction only has meaning when we apply current political boundaries. Only the occasional sea level rises that separated Africa from Eurasia temporarily restricted movement and the apes of the day exploited the opportunities that came and went like the ebb and flow of the tide.

The apes that made it across into the Eurasian part of the supercontinent thrived there in the mild climate that lasted from 14 to 9 million years ago (Figure 2). The absence of fossil apes on the African side from this time led to the suggestion that they went extinct there and that the ancestor of the modern apes, and by extension humans, re-entered Africa at a later date from the Eurasian side.[11] This claim was considered important because, if true, it would have meant that our own ancestors had originated outside Africa and dispersed back there some time later. The theory went on to suggest that, as climate deteriorated after 9 million years ago the apes became extinct except in the milder tropics which they had reinvaded from Eurasia.

The counterclaim to this theory came from those who argued that the lack of fossil apes in Africa did not necessarily mean that they were not there but that they had not been found. A number of discoveries of isolated fragments and teeth certainly seemed to suggest that the apes had not disappeared completely from Africa. Then came two reports, published towards the end of 2007, that seemed to vindicate those who argued for a continued presence of apes in Africa: a large species of great ape, possibly the ancestor of the gorilla, from Ethiopia and dated to between 10 and 10.5 million years ago;[12] and a second species, thought to be close to the last common ancestor of living African apes and humans, from Kenya and dating to between 9.8 and 9.9 million years ago.[13] This latter ape was very similar to one from Greece (dated to between 9.6 and 8.7 million years ago) that had been a strong candidate for the ancestor of living great apes and humans until then.[14]

One aspect that came strongly across from these studies was that the African apes were occupying similar habitats to the European ones, especially seasonal evergreen woodlands dominated by hard leaved (sclerophyllous) trees. As climate deteriorated after 9 million years ago, these forests disappeared from northerly latitudes and the apes went with them, remaining only in parts of tropical Africa and south-east Asia.

The debate on the relationship between African and Eurasian apes in the Miocene resembles others that we will meet later in this book when looking at their human descendants. Much of the confusion stems from the artificial division of Africa and Eurasia. If instead we consider the Afro-Eurasian land mass as a single supercontinent then the entire perspective changes and we get a much clearer picture. The canvas is made up of vegetation belts and sea barriers and the artist is climate. In this case the actors are the apes.

To sum up this particular episode, from the beginning of the Miocene through to the climatic optimum in the early part of the Middle Miocene, that is the period from 23 to 14 million years ago, tropical and subtropical forests dominated the landscapes of the supercontinent. Once the sea barrier that had separated Africa from Eurasia was lifted, the subtropical forests on either side were open to invasion by animals that could live in them but which had until that moment been confined to one or other side. In the case of the apes the expansion took a little longer because there was no ape capable of exploiting the seasonal forests when the barrier first disappeared. Climate had slowly started to change the forests of tropical Africa, reducing the rainforests with tall canopies and replacing them with seasonal woodland, and new forms of ape capable of living in such new environments appeared on the scene. It was these new apes that were able to expand across the huge tracts of seasonal forests. These changes in species range were simply the natural consequence of population growth in favourable situations and decline in bad times.

The last part of this story is almost a mirror image of the start. It ends in tropical Africa because that is where the once ubiquitous forests were left as climate took a downward spiral. These apes were not living everywhere across Africa but only in favourable habitats. Initially that had allowed them to live from tropical East Africa north into the Levant and from there west to the Iberian Peninsula and east to China. Their fortunes waxed and waned as climate changed the size and shape of this core area. In the end only two regions, one in tropical Africa and another in south-east Asia, were suitable for these apes and it is in one of these that the story moved to a new chapter.

We have come a long way from the distant asteroid impact of 65 million years ago to the African apes of 10 million years ago. Yet, despite the 55-million-year marathon, we are still unable to find an animal on the planet that resembles one that we could, by any stretch of the imagination, call human. We will have to wait a little longer but the major elements that will be critical to the human story have already made an appearance during this prelude. It will be helpful to us to have these fundamentals close to hand as we unravel the events that would one distant day produce the writer and the readers of this book.

The story unfolds in the theatre that is our planet. It is a theatre with several interconnecting stages. The doors between some of the stages are sometimes closed, preventing actors moving between them; some stages are further away from the rest and are harder to reach while some have the doors closed for longer than others. At first the acts are restricted to parts of Africa and Eurasia but eventually Australia and then the Americas are brought into the play. The scenes and stage sets change with each act and on each stage. The stage manager is climate, constantly changing and rearranging the scenes.

The human story unfolds against a backdrop of deteriorating climate. The main trend was towards a cooler and more climatically unstable planet. We have already seen evidence of this and it is about to get much worse. There were temporary reversals to the trend. Some were of long duration, like the climatic warming of the early Miocene that lasted for around 8 million years; others, at 55, 34, and 23 million years ago, were in the order of hundreds of thousands of years. There would undoubtedly have been many other, even shorter, periods of warming but they were all embedded within the big picture of climatic cooling. The world changed as a result. Polar ice sheets appeared and the tropical and subtropical vegetation belts that once reached Siberia contracted to suitable climatic areas at lower latitudes. The story of the primates shows us how animals responded to such drastic changes.

The key point of my argument is the one that makes, for me, our story such a beautiful one. It is the role of chance. It is about how unexpected events and situations altered the course of the story in unpredictable and unforeseen ways. We have seen examples of this with the asteroid impact at the K/T boundary that opened an unexpected window for our ancestors. Chance is everywhere in our story and it has affected it in subtle, as well as dramatic, ways. The successful apes that ventured into the seasonal subtropical forests of Eurasia could not have done so, despite their newly acquired taste for nuts, had a land bridge not formed in Arabia that enabled them to get across. Without the land bridge, then quite possibly another kind of animal from within Eurasia would have found ways of exploiting these foods and habitats, or it might all have remained untapped.

The cast of actors in the play is huge and it changes through time. Some actors disappear while new ones are introduced into later acts. Others remain throughout but we may not readily recognize them as they change their appearance in myriad ways. The cast is, of course, made up of the animals of the planet. The lead roles are taken up by the primates, not because they are any better than the rest but because this play is about them. Some other animals may sometimes get key parts while most contribute to the acts and scenes as extras.

I will separate the lead roles into two: conservatives and innovators. The conservatives, as we would expect, do not like to change roles at all and do their utmost throughout the play to stay the same. The innovators, on the other hand, have the ability to reinvent themselves over and over again. Quite often it is not through choice but because if they had not changed they would have disappeared from the theatre. They do not change consciously, as no actor knows what lies ahead in its future. When the future is more of the same, then the conservatives do well but when it is continuously changing in unexpected ways, then some lucky innovators do best while most others vanish along with the conservatives.

Conservatives and innovators are part of a continuum in our story. Innovators are always born out of conservative parents and their children will often settle into new ways and strive to become conservatives themselves. The reason is that, unaware of the future, the focus will always be on matching the present in the closest possible way. It is this that, when the scenes change suddenly and unexpectedly, can often lead to their downfall.

Conservatives responded to climate and environmental change by following their preferred stage sets across the theatre. Sometimes, the sets were thrown out of the theatre altogether, and the conservatives went with them, never to return. When climate favoured the expansion of suitable habitat then populations grew and spread. This was the case for the early tree-dwelling primates of 55 million years ago that managed to occupy huge areas of forest from southern Asia, across to North America, and into Europe via Greenland. When climate cooled and the evergreen forests vanished from higher latitudes many species went extinct. Others went locally extinct in these latitudes but managed to survive in refuges of suitable habitat nearer the equator. In most cases range contraction involved extinction of local populations and not, as often thought, migration of populations into refugia, which were often already occupied by other populations of the same species.

It is enormously important to our story that we should be clear in our understanding about how the geographical range of any species can move around different parts of the planet through time. I will use a well-documented example to show what usually happens. The collared dove is a familiar sight to many city dwellers across Europe. It is a bird that is at home in parks and gardens where it is very successful. A hundred years ago the collared dove was a rarity on the continent. This southern Asian bird gradually spread into Turkey and, from there, north-westwards right up to the British Isles and south into the Iberian Peninsula.[15] Nobody is really sure what triggered the expansion but the doves' success in exploiting parks and gardens, typical of urban and suburban Europe, must have played a role. In a sense humans had created a new habitat, just as climate does, and this particular bird like our early tree-dwelling primates, moved in.

Now, nobody has ever claimed to have seen collared doves reaching Britain in droves from Turkey. It did not happen that way. What happened was that collared doves settled in suitable areas in south-eastern Europe, where they were well established by 1900, and reproduced successfully. Their offspring could not stay where the parents lived, so they moved a kilometre or two down the road to the next park. Like this, kilometre by kilometre, the birds got across Europe. In Britain, the first pair bred in Norfolk in 1955 and the population had expanded to 19,000 by 1964. There are in the order of several hundred thousand today and the European population is estimated at 7 million breeding pairs. Despite a detailed knowledge of this expansion, its causes remain obscure. That should be a lesson to us when attempting to understand events that took place tens and hundreds of thousands, even millions, of years ago with a meagre knowledge based on scattered fossils and artefacts.

There was no migration of collared doves; it was simply a demographically triggered geographical expansion and it happened in less than a century. If we were to go back into prehistory, with the weak time resolution of the archaeological record, we simply would not see this gradual change that took under a hundred years to unfold. Instead we might, if we were very lucky, find in an imaginary cave, an archaeological level without any dove bones followed by another with lots of bones. Let us transpose the example to a human one that will occupy us later in the book: the archaeological record seems to indicate that our Ancestors were living in north-east Africa around 60 thousand years ago; by around 50 thousand years ago, if not sooner, they started to spread eastwards and reached Australia. The distance is admittedly substantial but so was the time elapsed. Let us compare those humans with the collared doves to see how each fare when it comes to speed of geographical expansion. It took the collared doves roughly 55 years to get across the 2,500 kilometres from Turkey to Norfolk, which is a rate of 45 kilometres per year. Now, the Ancestors living in Ethiopia 50 thousand years ago were approximately 15,500 kilometres away from Lake Mungo in south-eastern Australia, where we have the first well-dated evidence of the Ancestors at 46–50 thousand years ago.[16] Let us assume they got there by 45 thousand years ago. That gives us a rate of just over 3 kilometres per year, quite poor in comparison with the collared doves. This is an unfair comparison though, because collared doves reproduce at a faster rate than humans. Their generation time is effectively one year, which means that their rate of spread was 45 kilometres per generation. If we take the human generation time at 20 years, then our calculation converts the rate of spread to 60 kilometres per generation, which is in a similar order of magnitude to the dove. These are admittedly rough calculations but they serve to illustrate one thing very clearly—that there was nothing particularly special about human geographical expansion in prehistory, and it most certainly was not a migration of peoples.

Individuals and populations of species that track suitable environments in which to live in the manner that I have just described belong to conservative species that change little by sticking to what they know best. When the speed or intensity of environmental change is too great for such a species to handle and most suitable habitats in which a living can be made disappear, then extinction is the most likely outcome. Asteroid impacts and some twenty-first-century human activities are extreme examples of such big-impact changes. Quite often, however, the rate or intensity of change is less severe and this allows at least a few populations of a species to survive in some part of the geographical range. These populations continue to do what they have always done. If conditions improve at a later date, there will be expansion into new areas; if they remain the same then the surviving populations will continue at the new stabilized level and if they deteriorate further extinction may follow anyway.

My interest here is with those that manage to survive in some part of the geographical range. These populations will be continuously adapting to their environment, which will be perceived as relatively stable compared to those which had to be abandoned. Provided conditions do not change too much, individuals best able to handle the situation will be favoured through natural selection. In this way, some animals can exist almost indefinitely. In a sense the great apes of the tropical and equatorial forests of south-east Asia and Africa would fit here. Many of the early ape versions actually died out but some managed to adapt and stick to the forest way of life to reach the present. Although they continued to evolve within this wooded context, of those who made the jump away from the forest only humans remain.

Crocodiles provide an excellent example of a way of life that has persisted for millions of years. These reptiles emerged in the Cretaceous (145–65 million years ago) and made it through the K/T boundary. They were once more widely distributed than today and lived in many parts of Europe but as tropical environments shrank, these animals became captives in their own homes; yet within the confines of their habitat prisons, they continued to successfully make a living as low budget large meat-eaters of warm waters. The long-lasting success of the crocodile body plan does not mean that crocodiles stopped evolving: on the contrary, crocodiles continued evolving, albeit within the bounds of a successful basic design. In fact, the Nile crocodile only made its first appearance in the fossil record of the late Pliocene, around 2 to 3 million years ago; and most modern crocodiles only appeared and spread geographically at this time, when many other, more conservative, kinds of crocodile were going extinct because they could not handle the conditions imposed by a cooling planet.[17] The point is that here we have a case of a good model that has gone through a series of versions without losing its basic design. In a sense we could consider the crocodile plan to have become specialized, and that restricted it to specific parts of the globe, but it nevertheless managed to survive in an increasingly inhospitable world.

Innovators live on the edge, in peripheral parts of the geographical range of the species where conditions are not ideal but are good enough to allow some to survive. In many cases these edge populations only persist because a surplus stock of individuals from the good areas continues to spill over. They are referred to as sink populations, constantly drawing on immigrants to keep them going. Why should these populations of below-average individuals matter to us? Let me illustrate why with an elegant example.

My colleague and long-time friend Larry Sawchuk of the University of Toronto has been studying the population of Gibraltar, a small British territory at the southernmost tip of Europe where I come from, for many years. He is an anthropologist with a particular interest in the impact of disease on humans. Gibraltar is a great laboratory because after it was captured by the British in 1704, detailed records were kept by the military of all persons residing in the place: new arrivals, departures, births, deaths, marriages. Nobody escaped the attention of the scribes who worked for the British Empire.

Gibraltar was not a very nice place in which to live during the Victorian era. Conditions of sanitation were bad, civilians lived under crowded conditions, and there was scarcity, particularly of drinking water.[18] Being in a Mediterranean climate, the shortage of water was exacerbated during the 3-month summer drought. People tried to resolve the problem by digging underground cisterns that would store winter rainwater. Lucky ones had access to the few wells that sucked out the water table. The poorer folk had neither wells nor cisterns. Larry was able to equate access to cisterns and wells with social and financial status. The poorest people had neither, the next along could access cisterns, others wells, and the richest had both.

Larry looked at the records for the period between 1873 and 1884. The annual rainfall was typically erratic: there were good years, when a lot of rainwater could be stored, and bad ones when very little water could be collected ahead of the summer. He equated the amount of rainfall of the previous winter to the level of stress the population was under. The driest years were clearly the most stressful. It was then that people were most likely to drink contaminated water. The quality of the food supply was also compromised, for example, through the adulteration of milk. It is easy to build a picture of the awful situation.

The detailed records allowed Larry to work out where children under the age of one year, many of whom died from weaning diar-rhoea, had lived. From a detailed house-by-house survey carried out in 1879 he could determine whether the child had come from a house with a cistern, well, both, or neither. His results were stunning. As we would expect, childhood mortality during normal conditions was highest among the poorer people who only had access to the worst water and it was lowest among the better-off ones who could get water from wells and cisterns. The crux came, though, when he looked at the bad years; it was then that severe drought limited access to good drinking water for many more people. Could we predict the results? I did not. What he found was that it was the poorest people who survived best of all under these conditions! These people were used to coping with the strain of having to survive drinking bad water all the time so when drought hit, they felt the effects least of all. As long as years were wet, wealthier people were fine, but as soon as things got bad they simply could not cope.

My example is not far off from what I think happened frequently and drove human evolution. I have called this the survival of the weakest,[19] clearly tongue in cheek but vividly bringing home the point that it was not always the strongest and best at surviving in particular situations that fared best in unpredictable and changing environments. Those that occupied core areas either went extinct or moved with their preferred environments when these contracted in size or changed geographical position in response to climate change. They were the conservatives. Those on the edge had to constantly adapt to variable conditions; they were jacks of all trades and could even stay put when conditions worsened. In fact if such conditions persisted it was these jacks (or innovators) that fared best, their numbers augmented and their geo-graphical range expanded. If there was genetic interchange with the core populations, then the innovators would gradually win over the conservatives. This would appear as change within the species. If, on the other hand, the innovators became genetically isolated from the conservatives, perhaps as climate inserted an ecological barrier between them, then the jacks would move forward as a new species and either the others would stay as they were or their populations would shrink in size or even disappear altogether.

The range of woodland habitats available to the early African Miocene apes that we met earlier in this chapter permitted a number of species to evolve in new directions. Each had its own peculiar diet and a number of them managed to break away from subsisting on the traditional diet of ripe fruits. We can see how edge populations of apes, perhaps excluded by other apes from the richest habitats and unable to regularly survive exclusively on such fruit, would have gradually found ways of eating other plant matter, from leaves to nuts. As these populations became isolated, any new changes to the teeth or the gut that allowed these edge apes to improve the digestion of the alternative foods were favourably selected.

It was the innovators that were able to turn disadvantage into success. The alternative foods became the staple of the new species while others remained on the original diet of ripe fruits. The novel foods allowed the new species to move into areas that the others, in the absence of fruit, had never been able to enter because there had simply been no way of making a living there. A change of behaviour, and later of anatomy, as a result of being kept on the margins of the best habitat by fellow competitors became an advantage. In the case of the Miocene apes, it was an advantage that allowed them to expand away from tropical Africa and exploit the huge areas of seasonal subtropical forest that occupied vast tracts of Africa and Eurasia.

We must not forget chance in all this. Chance does not only come into the scenes of the play to affect the actors. Chance also influences the actors themselves in a very direct way. The apes that ran, four-footed, across the branches of trees in the forests around 20 million years ago evolved flexible joints that must have made their antics in the trees a joy to watch. Who could have predicted that many millions of years later, such flexible joints would come in handy to some of their descendants who would live, two-footed, on the ground, and would have a brain capable of imagining how to make a weapon from stone? In evolution, successful formulae match unknown futures. Over millions of years of life on Earth, they have been the minority.

1

The Road to Extinction Is Paved with Good Intentions

A ROUND half a million years ago a clan of people lived in some of the valleys of northern Spain, close to today's cathedral city of Burgos. To all intents and purposes they were recognizably human. They were intelligent, tall, and well built: they averaged 1.75 metres in stature, weighed around 95 kilos, had brains of comparable size to ours, lived in social groups, and were probably able to speak.[1] More than 5,000 human fossils belonging to at least 28 individuals have now been recovered in the Sima de los Huesos (the Pit of the Bones), a shaft within a cave in the hills of Atapuerca, and it has been estimated that they represent 90% of all the known human fossils from this period (Figure 3).

How did the bones get there? That is a question that remains cloaked in uncertainty and, like so many of prehistory's enigmas, wrapped up

Figure 3 Excavating a skull of *Homo heidelbergensis* inside the Pit of the Bones, Atapuerca (Spain)

Photo credit: Javier Trueba / Madrid Scientific Films.

in controversy. One group of scientists working here believe that the large numbers of human fossils, in the virtual absence of other animal remains except those of cave bears, is proof that this was not a place in which people sheltered and brought the animals that they had hunted. It was, instead, a place where they buried their dead, proof of the complexity of their behaviour and of self-awareness. The discovery, in 1998, of a beautifully-carved hand-axe among the human remains supposedly added weight to the argument as it suggested that this was a special implement that had formed part of the burial ritual. It was the only implement found in the pit and it was made of a red quartzite rock unknown in the caves of the area. The discoverers called it Excalibur. To me the logic behind the claim reveals the extent to which some scientists are prepared to romanticize and delude themselves, when all they have are glimpses of what must have been a complex past. Can we be comfortable with the idea that a single stone implement really tells us that much about the way of life of a group of people that lived so far back in time?

Others were not convinced that this was a place of burial and argued that many bones had the marks of the teeth of carnivorous animals that may have dragged the corpses from the outside into the shelter of the pit. I do not know what the cause of the accumulation of fossils at the Sima was but I am grateful that luck preserved this fine collection for the following half a million or so years, allowing us the chance to argue over how they got there.

In the previous chapter, we paused around 9 million years ago, when apes were managing to survive in the remnant forests of the tropics. I have started this chapter eight and a half million years later with creatures recognizably human. Before getting too wrapped up with these people I would like to make a swift expedition, in this chapter and the next, into the intervening eight or so million years to give us some understanding of how the people of the Pit of the Bones got to be there in the first place.

Each time that the scientific literature publishes a new fossil ape, proto-human,[2] or human, the picture of our evolution appears to become increasingly complex and difficult to comprehend. This is because we have only a few, usually incomplete, specimens available for study and this inevitably leads to much speculation about the relationships between them. It is like pretending to know what a 10-thousand-piece jigsaw puzzle looks like from only 100 pieces. Often we end up with colourful evolutionary trees that somehow connect the different fossils right up to us. These interpretations then get translated by the popular media and find their way into magazines and television documentaries as undisputed facts.

The Pit of the Bones sample is uniquely large and allows us to get a handle on the range of variation among people that lived a long time ago. We have nothing comparable between 9 million years ago and the Pit of the Bones, so our reconstruction of events must be tentative and cautious. Rather than get too involved in a discussion of which fossil might be the most likely candidate to be our Ancestor, I will instead identify the kinds of proto-humans and humans within broad categories and periods of time. The entire period can be conveniently classified into three blocks that neatly correspond to geological time zones: the Late Miocene, between 11.6 and 5.33 million years ago; the Pliocene, between 5.33 and 1.8 million years ago; and the Early to Middle Pleistocene, between 1.8 and 0.5 million years ago (Table 1).

The period between 11.6 and 5.33 million years ago is significant because it was the time when our limb of the ape family started to peel off from the branches that would head in the direction of the gorilla and the chimpanzees. We had already parted ways with the orang-utan side of the family, which had by then gone down its own independent evolutionary course in the forests of south-east Asia, and the gorilla clan was next in line for the separation.[3] Although most recent estimates have put the gorilla–human divergence around 8 million years ago, as we saw in the previous chapter fossilized teeth of an early gorilla, dated to around 10 million years ago, have recently been discovered in Ethiopia. If further evidence confirms these findings it will mean that the gorilla line may have separated much earlier than previously supposed and probably around 11 million years ago.

Common name used in this book	Latin name	Time span (millions of years ago)	Geographical distribution
Late Miocene (11.6–5.33 million years ago)			
Toumaï	Sahelanthropus tchadensis	7–6	Chad
Millennium Man	Orrorin tugenensis	6.1–5.72	Kenya
Kadabba	Ardipithecus kadabba	5.77–5.54	Ethiopia
Pliocene (5.33–1.8 million years ago)			
Ramidus	Ardipithecus ramidus	4.51–4.32	Ethiopia
Lake Man	Australopithecus anamensis	4.2–3.9	Ethiopia, Kenya
Lucy	Australopithecus afarensis	3.9–3.0	Ethiopia, Kenya, Tanzania
Abel	Australopithecus bahrelghazali	3.5–3.0	Chad
Pre-Flores Man	? Australopithecus floresiensis	?	?Southern Asia
Flat Face	Kenyanthropus platyops	3.5–3.2	Kenya
Taung Child	Australopithecus africanus	3.3–2.3	South Africa
	Paranthropus aethiopicus	2.8–2.3	Ethiopia, Kenya
	Australopithecus garhi	2.5	Ethiopia
	Paranthropus boisei	2.5–1.8	Malawi, Tanzania, Kenya, Ethiopia
	Paranthropus robustus	2.0–1.8	South Africa
Handy Man	Australopithecus habilis	2.33–1.8	Ethiopia, Kenya, Tanzania, South Africa
Lake Rudolf Man	Australopithecus rudolfensis	1.9–1.8	Kenya
Early Pleistocene (1.8–0.78 million years ago)			
Handy Man	Paranthropus boisei	1.8–1.4	Malawi, Tanzania, Kenya, Ethiopia
	Paranthropus robustus	1.8–1.5	South Africa
Handy Man	Homo habilis	1.8–1.44	Ethiopia, Kenya, Tanzania, South Africa
Lake Rudolf Man	Homo rudolfensis	1.8–1.4	Kenya
Georgian Man	Homo georgicus	1.77	Georgia

Table 1. Currently identified species of proto-humans with approximate time span and geographical distribution

This table includes a hypothetical ancestor of Flores Man and tentatively places Handy Man, Lake Rudolf Man, and Georgian Man in the genus *Homo*.

The most recent estimates of the chimpanzee–human divergence put this after 5 million years ago, some bringing it down to as recently as 4 million years ago. A feature of the split is that it seems to have taken as much as 4 million years to complete and this has led to the controversial suggestion that some time after the initial separation, the human and chimpanzee lines mixed their genes only to part again some time later. The alternative, simpler, explanation appears to be that there was a large ancestral population, which could have been around 50–75,000 individuals,[4] that slowly split into two.

The genes of living apes and humans, despite the imperfections of the science, are telling us that the first split among the ancient apes separated the lineage of the orang-utan and that happened before 9 million years ago. Next came the gorilla line which probably split around 8 million years ago, although recent fossils seem to suggest it may have been earlier than previously thought. Finally came the human–chimpanzee separation, and that was somewhere around the 5-million-year mark. So, important things seem to have been happening in the first of our three periods but are there any fossils that might make the picture a little bit clearer?

We have three fossil species from this period. The earliest has been nicknamed Toumaï (*Sahelanthropus tchadensis*),[5] which is a name that the people of Chad, in the Sahara Desert, give to children born at the start of the harsh, life-threatening, dry season. Toumaï literally means 'hope of life' in the Goran language of these people of central Africa, where the fossils have been found. Nine specimens have been described comprising a skull, fragments of jaw, and a few teeth. Toumaï lived along the shores of a lake, some time between 6 and 7 million years ago, and may have walked upright, but this is not certain. The skull, holding a brain the size of a chimpanzee's, seems to combine features of an ancient ape with others that seem to anticipate later proto-humans. Needless to say the discovery generated a fiery controversy among those who, at one extreme, saw Toumaï as an ancient species on the direct line to humans and others who dismissed it as an early kind of gorilla. Toumaï's 6 to 7 million-year-old date tag presents us with a problem as the date is somewhere before the predicted split between humans and chimpanzees, around 5 million years ago. This leaves us with two options: if Toumaï is in the direct human ancestry, then the molecular clock estimates of the time of the human–chimpanzee separation are too young; if, on the other hand, the molecular clock estimates are correct then Toumaï lived some time before the partition and cannot belong to our direct, post-chimpanzee split, ancestry.

Thirteen fossils of the second contender to the throne of earliest and exclusive ancestor to the human lineage were initially recovered from the Tugen Hills, Kenya, in 2000.[6] In all, 22 fossils belonging to six individuals have now been found. The species was nicknamed Millennium Man (*Orrorin tugenensis*), although the scientific name it was given means 'original man' in the Tugen language. It lived some time between 6.1 and 5.72 million years ago, around a million years after Toumaï. Although closer to the molecular calculation of human–chimpanzee divergence than Toumaï, Millennium Man is still older than the estimated time of the split and presents a similar conundrum. In contrast to Toumaï, no skull has been found but instead several leg bones, from the thigh, have been recovered. These femurs give us clues to how Millennium Man moved. The scientists studying Millennium Man claim that it walked upright, a key trait of the human lineage, and have taken the argument further by adding that its gait and walk were closer to humans than later species that had, until then, been assumed to have been the ancestors of humans.[7] Although most would accept that Millennium Man walked upright, fewer are convinced that it moved in a human way, or that it represents a direct ancestor of later humans. Once again, meagre data open up opportunities for radical claims and counterarguments that do little to advance our understanding.

If the skull was Toumaï's strong point and the thigh Millennium Man's trump card, then it was the teeth that made the third aspirant, who entered the fray in 2001, stand out. Eleven fossils, representing five individuals, were reported from the Middle Awash area in Ethiopia and they had lived there some time between 5.8 and 5.2 million years ago. A year later, six new teeth were discovered and the time frame was narrowed to between 5.77 and 5.54 million years ago. Although these creatures were more recent, it is possible that they could have overlapped in time with Millennium Man, who lived 1200 kilometres further south. We will refer to this third candidate as Kadabba (*Ardipithecus kadabba*).[8]

Let us stop for a moment and think of what all this means. For the long period between 7 million and 5.54 million years ago, we have a handful of partial specimens from three tropical African sites. For some we have a head but no body, for others legs but no head, and for others mainly jaws and teeth. In all, they amount to fewer than twenty individual beings that lived their anonymous lives across this vast land at some point within a one-and-a-half-million-year period of the remote past. You would think that we would treat these few and incomplete finds with some sense of humility and proportion, but no, despite the dearth of material, the three candidates are hotly contested as our ancestors.

It seems that this is not enough for some. To further support the claim, each nominee has been given its own distinct status in a genus of its own: not only are they seen as singular biological entities, they are so different from each other that they must merit elevation to a higher order category of scientific classification. The truth is that we have no clue as to the natural range of variation in the characters that have been measured in these few specimens. It is perfectly plausible that our Three Amigos (as I will collectively call them), Toumaï, Millennium Man, and Kadabba, may have all belonged to the same genus or even the same species. All three could have been ancestors of ours but maybe none of them were.

We should not despair, however. We have found small, chimpanzee-sized apes that appear to be able to walk upright and the time that they are around broadly coincides with the time that we think proto-humans and proto-chimpanzees parted ways. That is an important step but can we find the trigger that set these changes in motion? That is never going to be easy when working with timescales in the order of hundreds of thousands to millions of years. In recent times our understanding of the geological and climatic changes affecting the areas of tropical Africa where the Three Amigos lived has improved considerably and this will allow me to paint the backdrop to this particular act of our play.

Two big geological events set the scene at around 8 million years ago. The first, at 8.5 million, was the start of a period when the water circulation from the Atlantic into the Mediterranean became restricted. At the time the connection between the eastern end of the Mediterranean Sea and the Indian Ocean had been shut, as Arabia had by then docked into its slot between India and Africa. The Atlantic flowed into the Mediterranean at its western end along two channels but as Africa kept on barging into Europe, land was being lifted and water flow was increasingly constrained by new barriers. Matters got progressively worse and parts of the Mediterranean were becoming brine lakes as more water evaporated than entered to replenish the losses. The salinity crisis started across the Mediterranean around 5.96 million years ago and hit its worst moment after all connection with the Atlantic was severed around 5.59 million years ago. Between 5.8 and 5.5 million years ago the sea level was reaching its lowest point.

The second event was a period of substantial rise of the Tibetan Plateau around 8 million years ago. The main result was that it kick-started the Asian monsoon. As summer heating caused warm air to rise above the plateau, moist air rushed in from the Indian Ocean to fill the gap. This moist air was forced up the mountain slopes and discharged as rain. The way this south-westerly monsoon works today means that southern Asia receives the bulk of the summer monsoon rain while north-east Africa remains arid.

But it was not always this way. The Earth wobbles in its orbit and the tilt on its axis changes in cycles of 19 to 23 thousand years, so the point when the Earth receives greatest heat from the sun shifts. When this insolation was strongest, it combined with the Tibetan land mass to generate the south-westerly monsoon. At other times, when insolation was at its weakest, it seems that it was not enough to produce this monsoon. A south-easterly monsoon developed instead, bringing rain from the Indian Ocean towards North Africa. This effect was enhanced because Africa was still moving northwards at this time, with the result that a large area of subtropical North Africa came into a position of the world that received the greatest summer insolation.

At this point two apparently isolated events, like the closing of the Strait of Gibraltar and the uplift of Tibet, came together to generate an unexpected change in the climate. By 5.8 million years ago the Mediterranean Sea was reaching its lowest point. Summer low pressure systems developed regularly over the former sea, enhancing the south-east monsoon over north-east Africa. Paradoxically, at the height of the salinity crisis, north-eastern Africa and the lands bordering the Mediterranean were in the middle of a very wet period.

The effects were felt most strongly along the coast and bordering lands of north-east Africa and the rains would have become less intense as we moved north-westwards away from the Indian Ocean's influence. Today, the south-west monsoon forces the moist Indian Ocean air up the Himalayas, where it discharges as rain. This monsoon water then makes its way down the slopes to feed major rivers and finally empty out into the Bay of Bengal, via the intricate network of channels that form the Ganges Delta. But what happened to the monsoon rainwater that fell on central and north-east Africa?

It is hard to imagine this area, most of it now under the grip of the Saharan sands, during a monsoon. The Nile and Lake Chad, in the central desert where Toumaï was discovered, are what remain of this monsoonal zenith. At its height the Sahara sucked in water into four huge basins that created massive inland seas of freshwater. These lakes then drained northwards and emptied in the area of the present eastern Mediterranean Basin, catapulting vertically as huge cataracts into the bottom of the salty Lake Cyrenaica. Together, the four lakes drained an area of 6.2 million square kilometres, an area roughly eleven times the size of Europe's largest country, France.

Toumaï lived along the shores of the ancient Lake Chad at a time when the area would have been much wetter than today. This was a rich environment interfacing gallery forest and savannah with frequently inundated land at one extreme and desert at the other.[9] The place was thriving with many kinds of freshwater fish and soft-shelled turtles, land tortoises, pythons and other snakes, lizards, and a wide range of mammals from large hyaenas and sabre-tooth cats to hippos, giraffes, antelopes, pigs, horses, and monkeys. The shores of Lake Chad were truly a mosaic of environments within a relatively small area, allowing animals to find many ways of making a living. Living in areas of localized ecological richness was the key to Toumaï's survival as it would be for many future generations of proto-humans and humans: the relationship between them and habitat mosaics is a recurrent theme that we can trace back to these simple beginnings.

Whether or not Toumaï was in our direct line of ancestry, it does nevertheless represent a kind of evolutionary testing that was coincident with far-reaching changes in the climate of this part of Africa. These experiments were starting to produce apes that, as the primordial rain forests shrank, were finding ways of living increasingly on the ground in wooded habitats where they could not live off eating fruit all year. Their teeth suggest they ate a range of plant matter, including underground roots, but some of them may also have been beachcombing the lake shores and getting an early taste for animal products. In a sense they were populations on the fringe of other apes and they were testing new-found ways of making a living. Toumaï and friends were innovators.

Returning to the shores of Lake Chad a small ape-like creature runs into cover and climbs up a tree to let a herd of elephants go by. These elephants look more like those we might see today on an African savannah, though not quite the same, and they are rather distinct from those Toumaï would have been used to seeing. The place is close to where Toumaï lived but we have taken a big leap forward, around 3 million years, in time.

The shores of Lake Chad between 3.5 and 3 million years ago were a mosaic of gallery forest and wooded savannah with open patches of grass, where horses, rhinos, and antelopes grazed. The animals were of similar types to those that Toumaï had known but they had changed in the intervening 3 million or so years. There was still water, permanent and seasonal streams, and the lake itself which was rich in fish, turtles, and crocodiles.

In 1993 a team of French scientists found part of a mandible and some teeth of a proto-human, which they named Abel (*Australopithecus bahrelghazali*) in honour of a deceased colleague.[10] They of course could not know that eight years later they would strike gold a second time and discover Toumaï. The 1993 find caused great excitement as it was the first of its kind to be recovered away from the South African–East African Rift Valley axis. It belonged to the genus *Australopithecus*, literally meaning southern apes, which had been known since 1924 when quarrymen in South Africa discovered the skull that came to be known as the Taung Child, who had lived some time after Abel.[11] Abel may have been a contemporary of the famous Lucy, the 40% complete skeleton of an adult female of around 25 years in age at the time of death, which was found in 1974. Lucy had lived in the area of present-day Hadar, Ethiopia, around 3.2 million years ago.[12] The Taung Child, Lucy, and Abel give us a good idea of the geographical range of these early small-brained proto-humans that were the dominant force throughout the second of our periods, between 5.33 and 1.8 million years ago.[13]

During this second period small-brained proto-humans were widely distributed across tropical East and Central Africa. They may have reached western areas of tropical Africa too but we have no fossils to prove it. They broke the tropical barrier and reached the southern tip of Africa but, surprisingly, we have no evidence of a similar northward expansion. It seems highly likely that these proto-humans would have spread to nearby areas north of Ethiopia, along the Rift Valley, and even into the Middle East, considering that they reached right down to South Africa which was twice the distance away. But did they?

We cannot tell how many species were involved over this long period because some of the fossils that have been identified as separate species may simply be geographical variants of one and the same. There are others that seem to be versions of the same type that replace each other as time goes by. The stem of this diverse and disparate assemblage seems to be Ramidus (*Ardipithecus ramidus*),[14] a descendant of Kadabba who lived in the same part of Ethiopia between 4.51 and 4.32 million years ago. For the next million years, all small-brained proto-humans seem to have been confined to tropical East Africa, from Ethiopia down to Tanzania, and it is only after 3.5 million years that we find them to the south and west.

What was happening to the climate of this period? It started off 5.33 million years ago with a dramatic and spectacular event. Just as the great rivers of the Sahara cut deeply into the ground as they drained the mega lakes into the salty basin of Lake Cyrenaica in the eastern Mediterranean, in the extreme west a river was also eroding the land. But this river was not draining a lake; instead it collected rainwater from the nearby wet Atlantic coastal areas. Slowly, the river cut back towards the largest body of water of the region—the Atlantic Ocean. One day it reached the level of the Atlantic, thousands of metres above the dry western Mediterranean Basin, and oceanic water began to trickle in. For 26 years it was a mere drip but as the new channel opened, it became a torrent and the world saw the birth of a super-waterfall that gushed water into the hot and dry abyss 3 kilometres below. In 10 years the entire western basin had reached the level of the Atlantic and it then poured over into Lake Cyrenaica to fill the eastern basin in a year. The new Mediterranean would alter the climate of Europe and northern Africa, making it more arid, and modern types of desert, semi-desert, and arid grasslands started to spread.

The landscapes were starting to look more like those of the present-day world and less like the warm, forested planet being left behind. Africa was still more wooded than today but the rainforests were shrinking in area and the woodlands were starting to break up. Ramidus, the earliest species of this period, seems to have continued the tradition of its ancestors and lived in a mosaic of environments dominated by woodland. What this means is that proto-humans were walking on the ground while still living in the forest; the old idea that walking on two feet started when our ancestors ventured away from the forest into the open savannahs no longer holds. It now looks likely that bipedal walking may have started on the trees themselves.

This startling conclusion has been reached from looking at the way that orang-utans walk.[15] Orang-utans share something with humans that gorillas and chimpanzees do not. All of them can stand upright but when chimpanzees and gorillas do so, the hind limbs are flexed. Orang-utans and humans, on the other hand, stand on straight hind limbs. This way of walking on tree branches gives the orang-utan great benefits. It can venture onto the flimsy branches on the outside of the tree crown on its hind legs, transferring the centre of gravity as necessary and hanging on, for safety, to other branches with the hands. It can then release one hand that can be stretched out to collect fruit that would not otherwise be accessible. The method of tree walking is also used to move between trees without having to come to ground. It seems that this is an ancestral form of locomotion present in the ancestor of all the great apes. It was retained by the orang-utans who kept to a similar lifestyle in the south-east Asian rainforests. The price was that, as rainforest shrank, they were left trapped in it. The orang-utan was probably the only conservative ape to have made it to today; all others either went extinct or changed their ways.

The African forests seem to have suffered more from climate change than those of south-east Asia and the continuous forest canopies were repeatedly broken up as forests opened. In Africa, forest fragmentation alternated with periods when the gallery forest canopy closed up again, with the appearance of humid woodland and with the recovery of the rainforest. It was an ephemeral and unpredictable situation. Canopy fragmentation would have limited the usefulness of the orang-utan style of canopy-to-canopy walking; new tricks, such as coming to ground and climbing up another tree to reach the canopy once again, had to be devised. We think of the gorilla and the chimpanzee as specialists of the forests that, like the orang-utans, kept on doing much the same as always while our wily ancestors moved onto the savannah. But they too survived by changing their ways.

What happened in Africa were experiments in a natural laboratory in which a wide range of tests at being successful apes were tried out. Many failed and were discarded. Only chimpanzees, gorillas, and humans made it to today. The chimpanzees and gorillas were at one end of the spectrum of experiments, deep in the forest; humans were at the other end, in the wooded savannahs. The chimpanzees and gorillas had to find efficient ways of getting around between canopy and ground, so they resorted to climbing vertically up and down tree trunks on all fours. Having adapted the skeleton to this kind of tree climbing, they dispensed with straight-limbed two-legged walking for ever. To get around between trees, they simply changed from portrait to landscape and literally climbed horizontally on the ground—they became knuckle-walkers. So knuckle-walking was a development and not the way that the ancestor of the chimpanzee walked. Neither did the ancestor of proto-humans.

For a long time it had been assumed that the knuckle-walking behaviour of chimpanzees and gorillas was an intermediate stage that proto-humans must have gone through in going from a life in the trees to one on the ground. Now, it seems that we have here another one of those cases in which an adaptation for a particular job came in handy, purely by chance, for another when circumstances changed. What some proto-humans did was simply to carry the tree-walking way of life onto the ground. Deeper in the forests knuckle-walking did a better job as much of the time was spent in the trees, but once down that route the knuckle-walkers condemned themselves to a forest existence; they were innovators turned conservative.

The proto-humans sacrificed access to the forest canopy by maximizing the upright way of life. Once fully committed to a permanent way of life on the ground, any modification that improved rapid bipedal walking and running would have been favoured, but the fossils suggest that this took a while. The small-brained proto-humans of this early part of our evolution appear to have retained long arms and other features that allowed them to return to the safety of the trees when they needed to. They were only gradually moving from the multilayered canopy forest into peripheral habitats in which they were experimenting, being innovators. In the process they were starting to focus on open habitats that no ape had previously lived in. Ramidus was already embarked on this particular adventure.

Ramidus vanished from the fossil record almost as quickly as it appeared around 4.4 million years ago, presenting us with the question of whether it went extinct or, instead, evolved into something else. Like its ancestor, Kadabba, Ramidus seems to have been confined to the Middle Awash Valley in Ethiopia. Shortly after its disappearance a new proto-human, Lake Man (*Australopithecus anamensis*), showed up in the same area, around 4.2 million years ago. First discovered south of Ethiopia on the shores of Lake Turkana in Kenya in 1994,[16] its geographical range was only found to include the Awash in 2006.[17]

These few observations give us a tantalizing glimpse that seems to show that Lake Man may have evolved from Ramidus and spread its geographical area of occupation southwards within a time interval of 200 thousand years. What made Lake Man so successful that allowed it to break away from the home in which the ancestors had been confined for over a million years? The answer is in its ecology. Ramidus seems to have lived, like its ancestor Kadabba, in woodland. This was not the multi-canopy rainforest of the proto-chimpanzees but instead grassy woodland, similar to more open, bushy, and grassland habitats. It was the landscape of the edge of the rainforest, where climate was encouraging an assault on the primaeval jungle. Here the climate was sub-humid and there was a dry season, so Ramidus was already living in a more stressful environment for an ape than Kadabba who, in comparison, had lived in relatively wet woodland. Ramidus was probably the first proto-human to venture into these peripheral habitats and exploit habitat mosaics in a limited way. Lake Man took it all a stage further.

Lake Man was no longer restricted to a narrow range of woodland habitats. We find him in dry, open woodland and bush, in places where gallery forests covered wide areas of floodplain; he also ventured into tree and shrub savannah and he lived close to sources of fresh water.

The climate was semi-arid and seasonal with annual rainfall in the range of 350 to 600 millimetres. Lake Man was living in a mosaic of habitats half-way between the more closed, cool, and humid woodland of its ancestors and the more open, warm, and dry wooded grassland that would dominate the future. That was lucky for Lake Man and his descendants.

Lake Man's wide habitat tolerance would have been favoured among ancestral populations, probably of Ramidus, on the margins of the wooded Ramidus heartland. It would have been a solution to living in the suboptimal rim—not being too fussy about where they ventured and taking what they could from each habitat. As climate changed and these edge habitats spread at the expense of the old woodland, many core Ramidus, specialized populations would have gone extinct but those on the margins found that they were kitted up to do very well in this new world. Like the poor folk of nineteenth-century Gibraltar whom we met in the Prologue, the stressed populations did best when the overall conditions worsened; and like the collared doves of the same chapter, they turned success into geographical expansion. The innovative Lake Man emerged in the borders of the woodland in which conservative proto-humans lived.

The later acts of the small-brained proto-human part of the play follow a remarkably similar pattern. Lucy (*Australopithecus afarensis*),[18] the famous skeleton found in 1974, belonged to a new proto-human that may have descended from Lake Man, appearing on the scene 3.9 million years ago just as we lose track of Lake Man. Lucy and her kind seem to have been even more adventurous than all their predecessors and we find them in environments more open than any occupied by those that came before them. They ventured even further, reaching south into present-day Tanzania where they walked over the famous site of Laetoli, discovered by Mary Leakey in 1978, leaving their footprints sealed in time. Laetoli confirmed that Lucy's clan walked upright, at least part of the time,[19] showing that terrestrial bipedal locomotion came ahead of brain enlargement and tool making among the proto-humans.

Lucy's people probably evolved in a similar fashion to the earlier proto-humans, living in more open grassland environment on the edges of others,[20] adapting to these situations and having the luck that climate continued to deteriorate and make these open savannahs and bush lands increasingly common. Walking upright seemed a good way to get around as the distances to be covered between trees got greater as the woodland opened up. The connection between proto-humans and woodland was not severed altogether but the degree of dependence on trees had changed. From the centre where all activities were carried out and food found, trees were now just places of refuge, to run to when threatened, to hide and spy on animals and neighbours, and from which to collect seasonal fruit.

This neat sequence from Lucy to humans may, however, be a red herring in our story according to some who would rather see Millennium Man on the right track towards humanity and Lucy's lot a mere offshoot, with Kadabba and Ramidus on the way towards becoming chimpanzees.[21] I do not agree with the second part of this interpretation which would require a reinvasion of the rainforest by the ancestor of the chimpanzee, at a time when the jungle was shrinking. Of the African great apes the chimpanzee, admittedly, has the widest habitat tolerance today and is able to live in wooded savannah as well as denser forest; but this is most likely the result of an expansion towards the more open woodlands from the rainforest rather than the other way round. It would seem to indicate that, during the long period of rainforest contraction, from 4 million years ago, there may have been several independent attempts by apes at colonizing the more open wooded environments.

That Lucy's line may not have been on the way to humans is another matter. It is quite possible that we are seeing one of several attempts at dealing with the increasingly arid, hostile, and open landscapes of East Africa. The Lucy proto-human production line may have ended with the extinction of one of her descendants.[22] Our own Ancestors might not have been involved with the Lucy experiment at all, having gone down a separate line instead. A third, less adventurous, design would have been the chimpanzee range. So there may well have been several trials at surviving in the plains of East Africa between 4 and 3 million years ago of which only two made it to the twenty-first century, as humans and chimpanzees.

If all this is correct and Lucy and her descendants[23] are out of the human picture altogether, are there any fossils in the period between 4 and 3 million years ago that we can pin down to our own ancestry? We do not know although there is one possible contender—Flat Face (*Kenyanthropus platyops*), a contemporary of Lucy's people; it is known from a 3.5-million-year-old cranium, jaws, and teeth excavated in 1998–9 and published in 2001.[24] The small teeth suggested a different diet from Lucy and the distinctive flat features of the face linked it to an enigmatic proto-human that lived in East Africa 1.9 million years ago.[25]

The remainder of the second of our three periods, from 3.5 million years onwards, saw a continuing deterioration of climate and the further break-up of forest into more open vegetation across much of Africa. Lucy's descendants reached their heyday with several species appearing on the scene and sorting themselves out across a range of open habitats, often close to freshwater and never far from the trees. Some developed robust bodies and teeth capable of dealing with nuts and tough, fibrous, plant matter. The geographical range broadened even further, into South Africa and west at least as far as Lake Chad. Some were still about at the end of the second period, the Pliocene, 1.8 million years ago.

The ancestors of humans are more cryptic in the fossil record than Lucy's clan are. Perhaps there were fewer of them than Lucy's descendants and they may have been living on the edge of a world of other proto-humans. Even more obscure are the proto-chimpanzees and proto-gorillas, probably because they were living in forests in which there has been little search for fossils. There is a lesson from all this that will serve us well later in this book: it is that there have been times in the human story when descendants of a common ancestor have branched off in different evolutionary directions and have found alternative ways of dealing with similar problems of survival. In the case of Pliocene Africa there was more than one way of being proto-human, just as there were at least two ways of being proto-chimpanzee.[26]

The Pleistocene, the main period of interest to us, started 1.8 million years ago with nothing like the kind of bang that the Gibraltar waterfall had produced at the start of the Pliocene. It was at 2.5 million years ago that a climatic threshold was reached when the planet came under the grip of large-scale climatic cycles that were to become the hallmark of the Pleistocene. The latter part of the Pliocene was already signalling the shape of things to come. Between 3 and 2 million years ago grassland habitats became more abundant, especially after 2.5 million years ago, but important areas of woodland survived, maintaining a spectrum of habitats from closed to open ones. It was also after 2.5 million years that the robust proto-humans, capable of dealing with tough plant foods, appeared in the wooded, grassy savannahs.[27]

This great diversity of proto-humans soon came to an end though, as open habitats overwhelmed the landscape of 2 million years ago. It left just a few tough types capable of living in this new hostile world. The pressures generated by a cooling, drying, and rapidly changing world had weeded out many designs but had also created fresh opportunities for innovation. Among these was an invention that would radically change the pattern of human evolution. It required an animal with a brain capable of imagining the final product of an operation before it had even been started, and the manual dexterity to implement it. Around 2.6 million years ago a small-brained proto-human made an implement out of stone and the world changed for ever.[28]

So who was there to welcome the arrival of the Pleistocene 1.8 million years ago? Some small-brained proto-humans made it and there was probably more than one species involved.[29] Some were clearly not our ancestors. Others seem to fit the bill. One was the enigmatic flat-faced Lake Rudolf Man (*Homo rudolfensis*), who some have linked back to the 3.5-million-year-old Flat Face. For some time *H. rudolfensis* had been regarded as an odd and distinct species, seemingly ahead of its time in brain volume and skull shape, and usually classified in the genus *Homo*: a skull found in 1972 in Koobi Fora, Kenya,[30] was thought to support the idea that it was the ancestor of later humans. It had a large, 750 cubic centimetres (cc), brain which stood out against earlier and contemporary proto-humans that had much smaller brains, ranging between 400 and 600 cc. But a computer reconstruction carried out in 2007 put the very existence of *H. rudolfensis* as a distinct species in doubt, suggesting that the skull, which had been recovered in a bad state, had not originally been put together properly. Not only did this give the skull its odd shape, it also meant that its brain volume had been overestimated. The new analysis recalibrated the brain down to 575 cc and put it firmly within the small-brained proto-human range; there was nothing special about it. Needless to say this new claim remains controversial, but it has cast a long shadow of doubt on *H. rudolfensis*.[31]

The other long-standing candidate for human ancestry was Handy Man (*Homo habilis*), so named because of a tenuous connection with stone tools found in the site where the first remains were discovered.[32] According to conventional wisdom, the small-brained *H. habilis* evolved into Upright Man (*Homo erectus*),[33] who was definitely in the human ancestry. The story was a neat one: the small-brained, chimpanzee-sized, *H. habilis* walked upright on the wooded savannahs of East Africa where it scavenged and where it found cover and protection among the trees. *Homo erectus* came later and was taller and had a much larger brain, which allowed it to venture further into the treeless plains where it hunted insatiably for meat.

This tidy sequence of events, from either *H. rudolfensis* or *H. habilis* towards a later *H. erectus*, sat increasingly uncomfortably in the eyes of those who found it hard to pin down these elusive small-brained proto-humans. When and where had they lived exactly and what features really defined them? New discoveries from Lake Turkana in Kenya, reported in 2007, confirmed these doubts and put the whole succession in question, in the process seemingly leaving our earliest ancestry in complete disarray.[34]

Even though nobody today seriously contemplates a direct relationship between brain size and intelligence, the increasing volume of the brain through time has been used as a proxy measure of our evolution. Our brains average around 1300–1500 cc, roughly twice the size of that of *H. habilis*. We are much bigger of course but, even scaling our brain volume to take account of our size, it is pretty clear that our brains are much larger in proportion. *Homo erectus* was the first to break the magical and arbitrary 1000 cc brain volume barrier; together with its tall stature and upright gait it was, to all intents and purposes, unquestionably human. The problem with dealing with averages is that they disregard the information contained in the variability found in any population. Take humans today. Even though brain volume may average around 1300–1500 cc, the range of variation is between 950 and 1800 cc. *Homo erectus*, with a range from 800 to 1030 cc,[35] had a smaller brain than us *on average* but some were already within our range.

What the latest Lake Turkana finds showed was that some *H. erectus* brains were quite small: the brain volume of an *H. erectus* who lived on the shores of the lake 1.55 million years ago was only 691 cc and within the range of *H. habilis*. This excited the scientists who reported the find, as it revealed to them that *H. erectus*'s brain volume variation was greater than had previously been realized. Why that should be so surprising is beyond my comprehension given the well-known variation among present humans that is, not unexpectedly, far greater than that reported for *H. erectus* which is based on a handful of specimens.

Much more interesting was the discovery in the same area of a specimen of *H. habilis*, who had lived there as recently as 1.44 million years ago. Since *H. habilis* and *H. erectus* first coincided in the fossil record around 1.9 million years ago, what the result showed was that they lived in the same area for close to half a million years, and this made the time sequence from *H. habilis* to *H. erectus* untenable. It is quite clear from this observation, as with that of earlier proto-humans, that the world was not averse to having more than one species around at the same time. Most, despite having invested in wonderful designs, had to face unpredictable challenges; sooner or later they found themselves in the wrong place at the wrong time and disappeared.

We have got used to being alone on the planet since time immemorial so we imagine that it was always this way. But there was always more than one way of being human, as this chapter has shown us.

2

Once We Were Not Alone

I N the previous chapter I suggested that it was strange that small-brained proto-humans had reached all the way down to South Africa from the Ethiopian heartland, but apparently had not made it into the Middle East which was so much closer. Until very recently the prevalent view was that it was *Homo erectus* who first managed to break out of Africa and disperse across Eurasia. We saw in the Prologue how the strict political division of continents, a distinction that has never existed other than in our minds, complicated our understanding of how early primates and apes got to where they did. The same simplistic distinction has been widely applied in the human origins debate. I think that this way of carving up the Afro-Eurasian land mass has held back progress in our understanding of what really happened, and how it happened, by more than two decades. We are only now just starting to recover from the negative effects of 'out-of-Africa migration' dogma and see human origins in a different light.

I was lecturing in Cambridge when the news of the discovery of the Hobbit (*Homo floresiensis*) was released in 2004.[1] The anthropological community was in shock as nobody could have predicted that humans of any description, other than us, could have managed to survive on the planet until so recently. Most people were trying to find an explanation. As usual, others sat on the fence to wait to see which way the wind would blow, while the predictable staunch minority was doing everything possible to debunk the new claim. When something does not fit a long-established scheme, the usual response is disbelief, which rapidly turns into ridicule.

The discovery seemed straight out of a Jules Verne novel. How could these little humans have survived in a remote island until so recently, and why were they so small? Most people seemed to be supporting the idea that *H. floresiensis* was the first human example of island dwarfism, a phenomenon long known in a host of animals. It often happens that when animals are cut off from the mainland and remain stranded on islands for many generations, the smaller individuals of the population are more successful than the large ones, presumably as they require less energy. Eventually, bizarrely small creatures that look like miniature versions of their mainland cousins can evolve. Among the most striking and well known are the dwarf elephants and hippos of some Mediterranean islands like Malta and Cyprus. This all seems to happen because resources are more limited on islands than on the mainland; so size reduction is a way of surviving in an impoverished island world. *Homo floresiensis*, it seemed, had gone the dwarf elephant way; and they were not the only ones because Flores, too, had had its own dwarf elephants.

I remember discussing *H. floresiensis* at the time. The idea of island dwarfism in humans did not sit comfortably with me. I found it hard to accept that, having achieved a large and complex brain with all its benefits, humans on islands should lose these gains because of reduced resources. Isolation on islands did not always have to end with dwarfs anyway. At the time we were already aware of the exciting discoveries of small-brained proto-humans (ascribed to a new species: *Homo georgicus*)

46

in Dmanisi, Republic of Georgia, which dated back to 1.77 million years ago,[2] so I wondered whether the Dmanisi and Flores people had some remote connection with each other. Did they, somehow, represent the spread of a proto-human population into Asia from Africa?

Meanwhile, back at home in Indonesia *H. floresiensis* was dragged into a custody battle, and abroad the debate about what it represented heated up and took a new twist. Now there were those who saw *H. floresiensis* as a pygmy human and not a new species at all, while others saw it as a person with a pathological condition known as microcephaly.[3] Further finds were published in 2005 by the original discoverers who now revealed the bones of several new *H. floresiensis* individuals; it was highly unlikely that all would have been pathological specimens so the latest bones seemed to close that particular chapter of *H. floresiensis*'s brief but animated history.[4]

I was in Australia a year and a half later when I heard an account of a study of the skeleton of *H. floresiensis* that had been published.[5] A detailed analysis that compared *H. floresiensis* with many different kinds of proto-humans and humans to try and see whether it was related to any of these was presented. *Homo floresiensis* came out as having a bit of a mixture of features: the skull resembled that of the African version of *H. erectus*,[6] but the rest of the skeleton was closest to one of the small-brained proto-humans—*Australopithecus garhi*.[7]

What did this mean? The authors of the study suggested three possible scenarios: either it was a new species that had started its career in Africa and ventured into south-east Asia prior to 2.5 million years ago;[8] or it emerged, in Flores or somewhere between there and East Africa, out of an earlier human population that had evolved its brain at a faster rate than the rest of the skeleton; or it was a species in the process of evolving from proto-human to human when it came out of Africa in which case it must have been prior to 2 million years ago, which is the time that we first observe early humans (*H. erectus*) anywhere.

It seemed clear that *H. floresiensis* fell well short of resembling us and so it was not a dwarf, pathological or ecological, version of *Homo sapiens*. In fact it was much older. Support came from a study of the wrist of *H. floresiensis* published in 2007.[9] It revealed a primitive wrist, not only very different from ours but also from that of Neanderthals. One fascinating implication of the change in the wrist is that it may have improved the manipulation of the hand in toolmaking. So, although *H. floresiensis* made tools out of stone, it probably did not have the dexterity of later humans. Unfortunately, we do not have similar bones from *H. erectus* so we cannot be sure when exactly the more-modern wrist first appeared; according to the authors of the study, it was likely to have been at some point between 1.8 million years ago and 800 thousand years ago.

Homo floresiensis seems to have lived a successful life on Flores, on current evidence until 12 thousand years ago.[10] We know that they butchered the carcasses of dwarf elephants and they controlled fire. It is tempting to think that these were ancient traditions passed on from generation to generation from the first ancestral *H. floresiensis* to settle on Flores but we have no evidence of this and it remains possible that they were acquired some time later. There is no evidence, however, of Ancestors on Flores before 10.5 thousand years ago so *H. floresiensis*'s behaviour could not have suffered from the Ancestors' influences. Critics have argued that the stone tools with which *H. floresiensis* had been linked were too complex to have been made by them and must have, instead, been struck by the Ancestors even though they had not overlapped in time. The argument is reminiscent of others that we will meet when we come to compare the Ancestors with the Neanderthals.

A detailed study of stone tool industries on Flores put paid to the criticism and showed that there was technological continuity in the way that the stone tools had been made, going back to at least 840 thousand years ago.[11] So fire, butchery, and the rest of *H. floresiensis*'s behavioural suite could equally be very old. It all pointed to *H. floresiensis* being the descendant of an ancient proto-human, which may have been widespread across many parts of the Afro-Asian supercontinent and got isolated and forgotten on the remote island of Flores. It also gave us a glimpse into the degree of complex behaviour of early proto-humans.

The most stunning and convincing evidence of sophisticated conduct came, however, from Dmanisi. A skull and jaw bone of one of the 1.77-million-year-old *H. georgicus* was of an individual that had lost all but one of its teeth several years before death.[12] Somehow it had managed to survive which seemed to indicate that care for members of the group was well developed in some proto-humans. Put together, all this new information meant that many of the traits that had been attributed in the past exclusively to humans, from the care of others to the making of tools and fire, were already present in proto-humans, not all of whom were necessarily our direct ancestors.

A great deal of fuss has been made about the earliest emergence of humans from Africa; but the evidence in support of a first geographical expansion from Africa into Asia by *H. erectus*, in his African version, is not well backed up by the available evidence.[13] The story has usually gone something like this: *Homo erectus* was the first human species that had long legs and a big brain, could make tools, and actively hunted in grassy savannahs for meat. This suite of features allowed it to migrate out of Africa and colonize Asia. Here we have another example of conventional wisdom that resists debunking even though the evidence in its favour is virtually nonexistent. Even worse, the logic behind the speculation (hypothesis would be too grand a term to apply) reveals a deeply ingrained misunderstanding of the way in which species expand their geographical ranges.

Let me take you back to the collared doves of the Prologue. They traversed Europe in under a century but it was not individual doves that migrated. Instead, it was a gradual expansion into new areas by the children and grandchildren of parent birds. The expansion was at the level of generations and not of individuals. So it would have been for the first proto-humans that would have gradually expanded into favourable habitats wherever these were. It would not have been a migration and I fail to see the relevance of, for example, long legs. What would have favoured expansion would have been reproductive output and suitability of habitats. The first proto-humans to expand from the geographical core area in tropical Africa did not have to wait until they were Olympic marathon champions to move out of their ancestral home.

They did not have to wait to be super-brainy either. How many different species of animals have traversed huge parts of the planet and occupied remote places? How many species of tree re-colonized former, distant, haunts after an Ice Age? All that was necessary was that their requirements matched the environments that they were moving into. The vast majority did so without particularly stunning brains, tools, or a body that made them superb at long-distance migration.

In the previous chapter we saw how Lucy's descendants, and probably other unrelated small-brained proto-humans, became abundant and widespread across open grassy woodland and savannah after 3.5 million years ago. At the time, the kinds of habitats that these proto-humans were thriving in were not confined to Africa; instead they stretched right across a mid-latitude belt from West Africa to China, with a southern prong down to South Africa.[14] Why these proto-humans should have been confined to areas of Africa, when there was so much suitable habitat beyond in Asia, has no logical answer.

Homo floresiensis and *H. georgicus* hint to us that they had not been confined to Africa at all but had, instead, spread far and wide. It is tempting to think that it was part of the same 3.5-million-year-old geographical expansion out from north-east Africa that not only took them south into South Africa and west to Lake Chad but also north into western Asia and east as far as Indonesia. For now, this remains conjecture although it looks increasingly likely that it was small-brained proto-humans that, some time between 3.5 and 1.77 million years ago (when we find them in Dmanisi), first ventured across the new grassy savannah habitats of Asia. They were well ahead of *H. erectus*.

The small-brained proto-humans never fully abandoned woodland as we have seen and they retained adaptations in the arms, for example, that would have served them well when they needed to seek cover in the trees. The kinds of grassy savannahs that stretched across these huge areas of Asia and Africa would have resembled open woodland rather than the treeless expanses of certain parts of central Asia today and would have made ideal habitat. The scale of these ancient habitats was brought home to me while I was researching for a lecture on human evolution and I turned to my favourite group of animals, birds, for support.

It all started with a colourful species of the crow family that I knew well from my fieldwork in south-western Iberia. The azure-winged magpie is today an inhabitant of open, savannah-like, grassy woodland. When the woodland gets too dense with trees the magpies are replaced by another crow, the jay. When the trees are too few they are replaced by the common magpie. Azure-winged magpies live in habitats that clearly resemble the ancient ones of the proto-humans. They are also found living in the Far East, in China, Korea, and Japan but are rather strikingly missing in the 10 thousand or so kilometres between there and south-western Iberia.

For a long time people could not believe that the magpies had once lived all the way from the shores of the eastern Atlantic to those of the western Pacific and so they came up with the theory that it had been the sixteenth-century Portuguese mariners who had brought the multicoloured birds back as pets. Some escaped and colonized south-western Iberia. That seemed fine until we discovered the remains of this bird in caves that had been occupied by Neanderthals in Gibraltar around 40 thousand years ago.[15] If they had lived in south-western Iberia then, one could no longer argue for a human-assisted introduction in historical times.

The DNA of living East Asian and Iberian birds has since been examined and the two populations seem to have separated some time between 3.35 and 1.04 million years ago.[16] This means that there must have been suitable azure-winged magpie habitat stretching all the way from Korea to Portugal around 3.5 million years ago (Figure 4). Some time between then and a million years ago as the climate became cooler and more arid, this habitat broke up to be replaced by treeless steppe and desert except in the moister western and eastern coastal fringes of Eurasia.

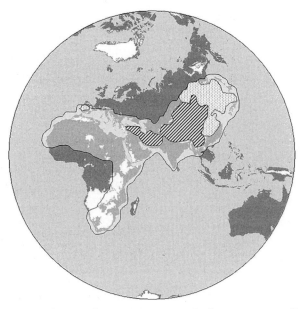

Figure 4 Extent of savannahs across Eurasia and Africa prior to 2 million years ago. Stippled areas are currently occupied by the azure-winged magpie. This species must have had a wider distribution that linked the disjunct populations at the start of the Pleistocene. Hatched areas are the central Asian mountains and the Tibetan Plateau

I decided to look at other birds that might have similar geographical distributions. I was amazed to find so many species that had populations scattered across this mid-latitude belt in what must have once been a continuous distribution from Iberia and Morocco in the west to China in the east. A staggering 40% of the breeding birds of Europe and western Asia seemed to fit into the pattern. Many were birds of wooded savannahs, like the magpie, but others were typical of open grassy plains and a few of woodland. Many were species of lakes and wetlands. The birds helped me form a picture in my mind of a vast belt of seasonal, semi-arid to sub-humid warm habitats, with some woodland and open plains and a preponderance of grassy savannahs and lakes. This was a lost world that had once spanned from the eastern

Atlantic to the western Pacific. The Dmanisi proto-humans of 1.77 million years ago were living in the thick of it. Dmanisi was the tip of the iceberg.

The Sahara, a vast area roughly equal to the United States, was the western end of this world. We have seen how large areas of it were taken up by huge lakes in the Miocene but wetlands came and went with alternating wet and dry periods long after this; Lake Mega Chad was active as recently as 5.5 thousand years ago and people thrived in this land of lakes.[17] Among the most exciting discoveries of recent times is one that reveals clearly the extent to which the Sahara was a huge wetland: in 1998 and 1999 relict populations of Nile crocodiles were discovered isolated in remote ponds in the mid-desert of Mauretania.[18] These, now dwarfed, crocodiles must have descended from Nile crocodiles that had lived right across the Sahara Desert at a time when it was dissected by interconnecting waterways, lakes, and marshes.

Looked at from the perspective of this vast belt of habitat, the presence of small-brained proto-humans from Ethiopia to South Africa, west at least to Lake Chad, north to Dmanisi, and probably east to Flores, makes perfect sense. It means that at the start of the Pleistocene 1.8 million years ago, and before the first appearance of the larger brained and taller *H. erectus*, proto-humans were already living across a wide area of the Afro-Eurasian supercontinent. The evidence available strongly favours a region in the core of this habitat belt, present-day Ethiopia, for the origin of the proto-humans, but was it also the heartland of *H. erectus*?

If we look closely at the available evidence for the first appearance of *H. erectus* in different parts of the world,[19] we can only reach one conclusion: it is that, on present evidence, we simply cannot say where *H. erectus* had its origin with any degree of confidence. We have *H. erectus* in East Africa 1.78 million years ago and in Java 1.81 million years ago,[20] and there is nothing resembling these anywhere in between. All other early sites that have been claimed to hold *H. erectus* either have fossils best classified as small-brained proto-humans or have been identified solely by the presence of tools. Since we know that proto-humans also made similar tools, we can have no certainty whatsoever that these sites actually represent *H. erectus*.[21]

Many books and articles illustrate the earliest expansion of *H. erectus* as starting in East Africa, with arrows pointing the routes out from this centre of origin. With the meagre data at our disposal, we should confine all these interpretations to the realm of fiction. The only thing that we can say with certainty is that *H. erectus* emerged, in all likelihood, from a population of small-brained proto-humans, somewhere in the grassy savannah belt that I have described. The close correspondence of early dates in areas as far apart as East Africa and Java shows us that *H. erectus* may have spread across a vast geographical area in a relatively short space of time. Such a rapid expansion is what would be expected when a new design first appears on the market and is successful; it makes it very difficult to locate the starting point of the expansion. If it is radically different from its predecessors and is able to exploit new situations or existing ones more efficiently, then the new form will spread rapidly. It is probably what happened with *H. erectus* and we will find an even clearer example when we look at the Ancestors.

The outcome of this part of the story is as follows: small-brained proto-humans now seem to have become widespread across the grassy savannahs of Afro-Eurasia some time after 3.5 million years ago. Some survived until at least 1.4 million years ago on the continent itself and if *H. floresiensis* is indeed the descendant of one of these populations, a few possibly kept on going on remote islands to a staggering 12 thousand years ago. *Homo erectus*, with his tall physique, first appeared with the start of the Pleistocene around 1.8 million years ago but we cannot be certain where he started his career or from whom he descended. His brain was larger than that of the small-brained proto-humans on average but there was considerable size overlap between the two.

Homo erectus and small-brained proto-humans survived side by side in continental areas, as in Lake Turkana in Kenya (Chapter 1), for up to half a million years: this shows us that the two managed to avoid competing with each other and that there was no immediately obvious superiority of *H. erectus* over the others. Like so much of the human story, there were often several kinds around at the same time and evolution was not a neat progression from one kind to another.

In the mid-1990s I spent many hours in the wilderness of the Coto Doñana in south-western Spain, a place reminiscent of the landscape that Neanderthals had lived in thousands of years earlier (Figure 5). The Victorian naturalist Abel Chapman had described it as a piece of Africa in Europe, because it was rich in herds of herbivorous mammals and its climate and vegetation were reminiscent of the semi-arid savannahs of East Africa. Doñana was sometimes a lush paradise and at other times a hostile desert. This place was the home of the Spanish imperial eagle, a majestic hunter at the top of the food chain. That was the popular image but the reality did not conform to the stereotype. During extended periods of drought the eagles did not bother wasting energy

Figure 5 The present-day landscape of sand dunes, pine woods and scrub, and lakes in the Doñana national park (Spain) most closely resembles the environments that Neanderthals occupied along large stretches of Mediterranean coastline

Photo credit: Clive Finlayson.

hunting. Instead they joined the vultures and scavenged the plentiful carcasses available.

The reason that I have digressed has much to do with the behaviour of the humans and proto-humans that we have been looking at so far. For decades people have been arguing about whether proto-humans and early humans hunted for meat or scavenged it. More recently the scavenging part of the argument has been divided into power scavenging, that is chasing the predators that made the kill away to secure the carcass, and passive scavenging.[22] These arguments and the elaborate analyses of the carcasses from fossil sites that have been used to support scavenging and hunting have been a source of concern to me for a long time. It seemed that if apparently full-time predators like our eagles, or indeed lions, scavenged when the opportunity arose and if supposedly committed scavengers like hyaenas hunted when they got the chance, then surely intelligent and opportunistic humans would have been capable of doing both things.

The scavenging–hunting debate has subsided somewhat in recent years but papers supporting one or other view still turn up in scientific journals from time to time. It is, in my opinion, a futile debate and one that reveals another of the flaws that seem to riddle the study of human evolution: it is the apparent need to generalize from limited and specific observations. Why should a perfectly good example of human hunting behaviour in some remote East African plain hundreds of thousands of years ago have to mean that all humans in that particular time frame hunted? Clearly it does not, and it does not even mean that the particular individuals who hunted in that very specific place and time always did so. We will meet some incredible examples of such overgeneralization when we come to the Neanderthals.

That the early proto-humans consumed meat, marrow, and fat from carcasses, no matter how it was obtained, seems incontrovertible. How much of their diet was meat is another matter altogether. Given the right conditions, large mammal bones preserve as fossils but plant and insect remains are more likely to decompose with the passage of time. The discovery of butchery sites, places where the bones of grazing animals have been found associated with stone tools, has probably biased us too far in the direction of the image of the meat-eating human. And the tools have taken us too far in the direction of stones, to the extent that we even define the major human archaeological periods as Stone Ages.

Sometimes a site is found that offers a glimpse of the alternative and diverse ways in which early humans exploited their environment: Gesher Benot Ya'aqov, a 780-thousand-year-old site in Israel, revealed a unique association of edible nuts with pitted hammers and anvils made by humans. The site was rich in wood and other plant material too, offering a tantalizing hint of the degree to which humans may have utilized and consumed plants.[23] The importance of mammal meat in the diet of prehistoric humans has been undoubtedly overemphasized, simply because bones survive better than wood or leaves.

In the past few paragraphs I have ventured away from the geography of proto-humans and humans towards meat consumption for a very particular reason. Meat in the diet of the early people has been given prominence as the key that enabled them to move away from Africa, in effect from tropical Africa. The role of meat has been implicated in many other ways, for example, being responsible for brain expansion in later humans.[24] So, to understand the first geographical expansions of proto-humans and humans we need to look at their diet.

Several things need to be cleared up to avoid confusion. Meat, as a food source for primates, is not exclusive to humans. Among the great apes chimpanzees regularly eat meat,[25] but the habit is not confined to them either. Savannah baboons regularly hunt animals up to the size of young gazelles.[26] Nor is the habit restricted to primates of the savannahs: an orang-utan, the classic fruit-eating great ape, has been observed eating meat from the carcass of a gibbon,[27] while the New World capuchin monkeys, a branch that separated from the Old World monkeys as far back as 40 million years ago, live in rain forest and are regular consumers of birds, bats, rodents, frogs, lizards, coatis, and squirrels.[28] None of these have needed big brains or technology to catch their prey.

It is quite likely that the proto-humans that started to live in habitat mosaics marginal to the rain forest already carried with them a general purpose diet that would have included an element of meat consumption. As with the ability to walk on two legs, an omnivorous diet was part of the forest survival strategy which would have worked well in more open habitats where fruits were scarcer and insects, frogs, reptiles, rodents, and larger mammals commoner and easier to find. When we compare chimpanzees to early humans, both regular meat-eaters, we find two very different strategies. From a common ancestor with a fairly unspecialized set of teeth, chimpanzees developed large canines and other teeth that may have been used to kill animals and consume meat while proto-humans kept a relatively unspecialized dentition but discovered tools which they used as substitutes for the large canines.[29]

This means that the proto-humans that managed to prise themselves from the forests into more open habitats must have had a general purpose diet and it was this flexible approach to food that made them successful. The expansion into non-tropical areas and open savannahs took place from 3.5 million years ago but we do not have evidence of tools until 2.6 million years ago. So, either the initial success of proto-humans had little to do with meat or it started off, with tools, a million years earlier than we think.

Proto-humans carried with them two things out of the forest: the abilities to walk on two feet and to eat a wide range of foods. Once in the open, technology simply improved the range of resources that could be taken, also allowing some foods to be processed more efficiently. Our own unspecialized dentition, digestive tract, legs, and use of gadgets reminds us that there is not that much in us that was not already present, in some shape or form, among the proto-humans. *Homo erectus* was an enhanced version of a proto-human. Those who followed *H. erectus* were similarly developments of the same theme.

The net result is that by 1.5 million years ago *H. erectus*, his descendants, and his regional variants lived across much of the Afro-Asian supercontinent and a new technology, characterized by beautifully carved hand axes, had made an appearance. The proto-humans were all but gone. By a million years ago the new tools had spread across this vast range. A million years later, in a land called Spain, archaeologists would name one of these hand axes Excalibur.

The past one million years have been marked by a radical change in climate in which 100-thousand-year cycles of climate variability, related to the changing shape of the orbit of the Earth around the sun, became the dominant force.[30] The cycle marked major global changes in climate, including the coming and going of the polar ice sheets (the 'Ice Age') and phases of wet and arid climate in tropical Africa.[31] It is the period when large regions of the northern hemisphere started to seriously suffer from the effects of the ice sheet advances and the woodland–savannah belt that stretched from Portugal to China finally broke up as steppes and deserts took over large areas. To the south, many areas of rain forest and tropical woodland broke up repeatedly in favour of open savannah and desert.

The period between a million years ago and the Pit of the Bones people of half a million years ago, with whom we started this excursion into our early evolution in the previous chapter, is fuzzy because of the poverty of fossils that can guide us. Not only do we have hardly any fossils but, to make matters worse, few are well dated. Despite the dearth of material, elaborate theories have been devised to reconstruct this murky past and new species of human have even been named from a smattering of skulls spread across a huge geographical area over a period of half a million years.

The conventional view may be summarized like this: *Homo erectus* had succeeded in establishing himself across Eurasia and Africa. Those in the Far East continued an independent line of evolution, in relative isolation, and may have survived until quite recently when they were overrun by a wave of Ancestors from Africa.[32] In the west, fossils which resemble *H. erectus* but have larger brains and a number of features that pre-empt the Ancestors first appear in Ethiopia around 600 thousand years ago.[33] These fossils are allocated to a new species, Heidelberg Man (*Homo heidelbergensis*), that allegedly spread out of Africa into Europe,[34]

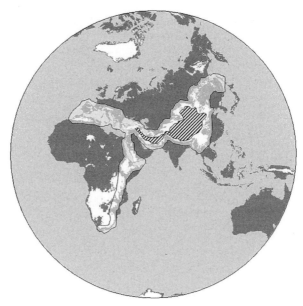

Figure 6 The mid-latitude belt (MLB) and its African extension, the home of *Homo heidelbergensis*. *Homo neanderthalensis* evolved in the MLB and *Homo sapiens* in the African extension. These areas were characterized by mixed topography which generated high ecological diversity. Hatched areas are the central Asian mountains and the Tibetan Plateau

and possibly as far east as China, around half a million years ago (Figure 6). According to this view *H. heidelbergensis* was the starting point of *H. sapiens* (Ancestors) and *H. neanderthalensis* (Neanderthals) (Figure 7).

Not everyone agrees with this view and some prefer to limit *H. heidelbergensis* to the European fossils that form the lineage from which only the Neanderthals would emerge.[35] A basic premise of both points of view is that *H. heidelbergensis* represents a new species that budded off from *H. erectus*. From that moment on *H. erectus* became relegated to a species, doomed to become extinct, that clung on for survival in remote East Asia. I do not agree with either of these views.

I am not concerned with what name we give each different kind of human or whether these constituted different species. I prefer to look

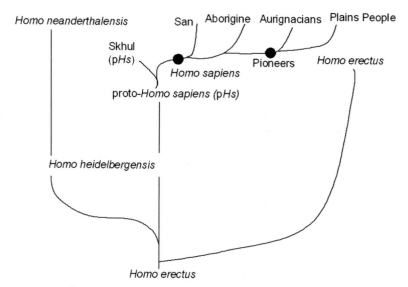

Figure 7 A basic human family tree as described in this book

at populations of humans and see, from the limited evidence at our disposal, what we can discern about their geographical distribution and their history. Up until the start of the Pleistocene 1.8 million years ago, proto-humans that may have reached non-tropical parts of Afro-Eurasia would have had to contend with seasonal climates but these would not have been as severe as those that would hit many regions during the Pleistocene. The conditions of the first 800 thousand years of the Pleistocene do not seem to have deteriorated sufficiently to prevent the expansion of populations of *H. erectus* across temperate latitudes but the onset of the 100-thousand-year cycles after a million years ago was a different matter.

It is after this that many populations of *H. erectus*, particularly but not exclusively, living away from the tropics would have begun to experience severe climatic downturns that would have been interspersed with periods of milder climate. The break-up of some of these populations, and the extinction of a number of them, would have become a real possibility. Although the extinction of populations of proto-humans and humans had been a feature of the early part of our evolution, the attrition caused by regular climatic cycles associated with the Ice Ages was a new phenomenon. The geographical expansion of *H. erectus*, beyond any boundaries ever reached by humans before, was rudely brought to a halt.

The mid-latitude belt, with its southern prong into East Africa, which had once been the subtropical woodland home of the Miocene apes and later the wooded savannah home of proto-humans and *H. erectus*, was now dissected by inhospitable deserts, treeless steppes, and snow-clad peaks. Like the azure-winged magpies, whose populations were severed, leaving relicts in Portugal, Spain, China, Korea, and Japan, many kinds of plants and animals suffered from recurring isolation. *Homo erectus* and his descendants would not have been immune to this turmoil.

This change in climate would have had two important effects on humans. On the one hand, populations would have become repeatedly isolated with the result that genetic differences would have started to appear between them. On the other hand, the stresses on the marginal populations would have intensified the need for innovation. Climate change had produced new kinds of apes in the Miocene, new kinds of proto-humans in the Pliocene, and probably *H. erectus* at the start of the Pleistocene. Now it was about to start to work on populations of *H. erectus* and it is no surprise that we detect an increase in brain size among these people at this time.[36]

Rather than try and lump all the fossils from this critical but shadowy period of our evolution into one or two species and lineages, I see them as regional variations on a theme. The theme is the adaptation of *H. erectus* to a rapidly changing world. Not all made it. Sometimes populations met and exchanged genes; at other times they were isolated from each other and went separate ways; and at other times some populations could have expanded into the geographical area of others and may have overrun them. Some of the fossils may have belonged to the precursors of *Homo neanderthalensis* or *Homo sapiens* but others may have belonged to lineages that later went extinct or merged with others after brief periods of isolation.

An analogy may help our understanding. *Homo erectus* went global. He was a corporation that became international and powerful. In time *H. erectus* franchises appeared in different parts of the world. Some broke away and became independent businesses; others, though they were profitable, had short-term difficulties with cash flow and went out of business; yet others were absorbed back into the *H. erectus* multinational.

Since the outcome of this period seems to have been two main products—Neanderthals and Ancestors (surviving relics of *H. erectus* and *H. floresiensis* apart)—we seem to have taken the naïve view that all fossils must have been on the line to either one or the other. We cannot be sure if all, only some, or none of these fossils were implicated in the next act of our story. *Homo heidelbergensis* probably came in different kinds, actually regional populations of *H. erectus* adapting and evolving to changing and varied circumstances imposed by the vicissitudes of the Middle Pleistocene climate.

Returning to the Pit of the Bones, we can now interpret these people as a local population in the Far West of the *H. erectus* range. They show features that suggest that they were part of a particular production line of *H. heidelbergensis*; one day *H. neanderthalensis* would emerge out of this line. They may have been part of a wider geographical population responding in a similar fashion to a particular environment. This population may have made it or it may have gone locally extinct. We may never know.

In the same hills of Atapuerca, an older site has provided fossils of a human (*Homo antecessor*) who lived there three-quarters of a million years before the Pit of the Bones People.[37] The fossils showed features found in *H. erectus* but they also hinted at things that we would find in *H. sapiens* much later. They were different from the later ones from the Pit of the Bones though, and there was no continuity between them.[38]

Atapuerca is a lesson for us: even within the same set of hills, people separated in time need not have been linked to each other and some vanished. Extinction must have been the fate of most Middle Pleistocene human populations. Our ancestors shared the planet with a number of other human forms. We were not alone.

3

Failed Experiments

I N 1932 human remains were excavated in the cave of es-Skhul (Mount Carmel). The finds were part of a joint expedition by the British School of Archaeology in Jerusalem and the American School of Prehistoric Research. Dorothy Garrod from Cambridge University was the director but she was away from the site when her assistant, T. D. McCown (from the American School), made the discoveries. Garrod had been the prize pupil of the great French prehistorian and priest L'Abbé Henri Breuil and had successfully excavated, at his instigation, the site of Devil's Tower in Gibraltar where she found the fragmented skull of a Neanderthal child in 1926. It was a decisive moment in her career and she soon turned her attention to the Middle East. Garrod was to become the first female professor at Cambridge, in 1939, holding the Disney Chair in Archaeology, although her appointment was not without its problems. The university's Vice

Chancellor noted that she would have to be an 'invisible' professor as the university's statutes did not take women into account. The situation was eventually changed after the Second World War, in 1948, by which time Garrod was already one of the leading pre-historians of her time.[1]

Garrod's work in the Middle East produced Neanderthal remains from the nearby site of Tabūn, also on Mount Carmel, and these discoveries opened up a chain of finds in the region during the course of the twentieth century, among them in Kebara close by, Jebel Qafzeh to the south, and Amud to the north-east. Some of the remains were clearly Neanderthal from the outset but others remained open to interpretation. Those from Skhul were thought to be 'modern humans' and distinct from the Tabūn Neanderthals. McCown and Keith, who described these fossils, changed their mind and later reinterpreted them as a single population 'in the throes of an evolutionary transition' in the direction of Neanderthals and Ancestors.[2] Since this time the position of the fossils from the Middle East has been in dispute, some seeing them as variations within a single population but others as two distinct species.

Many now regard the Skhul and the Jebel Qafzeh fossils as representing early, or archaic, modern humans. The terminology reveals the degree to which the application of taxonomic rules has placed a straightjacket on the understanding of the evolutionary process which does not recognize boundaries in time. What is, paradox aside, an archaic modern human? Is it archaic or is it modern? Even a cursory glance at these fossils by an untrained eye reveals that they were quite different from people today. The Skhul and Qafzeh crowd were a robust, large, and tough lot. Anatomically they are claimed to have features that place them on an evolutionary line that will lead to us and it is for this reason that they are called modern humans. But they also have many features typical of some of the *Homo heidelbergensis* of the Middle Pleistocene: they are evolving into modern humans and must be qualified as archaic. This is an unsatisfactory state of affairs. In this book I will refer to them simply as proto-Ancestors and the more recent ones as Ancestors. I will avoid the term modern as much as possible.

When were the Skhul and Qafzeh people around? Would knowing this help us in our understanding? The past two decades have seen an active effort to try and establish when the people of Skhul and Qafzeh lived in the Middle East; a similar effort has attempted to establish when the Neanderthals were also in the region. Part of the aim has been to see whether the two lived alongside each other. If this could be shown, then we would be looking at periods of coexistence. The problem from the start has been the lack of resolution of the technology available to us. We now have a handle on when these people were around but we cannot say with any degree of certainty whether they stared each other in the eye.

Combining the results carried out by various groups using different dating techniques points to a broad time period, between 100 and 130 thousand years ago, during which time the Neanderthals of Tabūn and the proto-Ancestors of Skhul and Qafzeh were living in the Middle East.[3] If we take a human generation to be twenty years, this time period represents 1,500 generations. Since this is the level of detail that we are able to achieve we simply cannot tell whether Neanderthals and proto-Ancestors lived side by side, or instead came and went at different times without meeting each other.

What we are able to conclude, which is relevant to this chapter, is that people with anatomical features that suggest they belonged to the *Homo sapiens* lineage were already living in the Middle East some time before 100 thousand years ago. Since we do not pick up Ancestors in Europe until 36 thousand years ago, the long time taken to reach this continent, which is literally round the corner from the Middle East, is in need of an explanation. We shall return to this point later.

The appearance of proto-Ancestors in the Middle East between 130 and 100 thousand years ago has usually been explained as the outcome of a northward geographical expansion of an African population at a time when the climate was warm and wet and savannahs expanded at the expense of the Sahara Desert. According to this view the Middle

East was connected by habitat to areas of tropical East Africa and the African fauna, including humans, followed the grasslands northwards. The Neanderthals, instead, were meant to have entered the region around 70 thousand years ago, from the north, during a cold and dry period. According to this view it was the harsh conditions in Europe that forced the Neanderthals south at a time when proto-Ancestors also retreated back into Africa.[4] It is a neat story but is it really what happened?

The period between 130 and 100 thousand years ago was one of relative warmth in the Middle East, coinciding with the global warming of the last interglacial period. It was a time sandwiched between Ice Ages but it would be wrong to assume that the climate was stable throughout this long period. There was a warm and wet period approximately 125 thousand years ago but this was followed by a warm and exceptionally dry spell, from 122 thousand years ago, and then another warm and wet period around 105 thousand years ago.[5] Since we know that there were proto-Ancestors and Neanderthals in the Middle East during this period which was by and large warm,[6] then the idea of Neanderthals entering the region from the north, after the proto-Ancestors had disappeared, during a period of severe cold climate seems highly improbable.

Instead, what we find are proto-Ancestors and Neanderthals living in the area when it was warm. The large errors in the dates of occupation do not allow us to say whether one was present when it was wet and the other when it was dry, or whether both occupied the area when it was either wet or dry. Is there any other evidence that might help us in trying to work out the climatic conditions when these people lived in the Middle East? Some researchers have turned to the animals that have been found fossilized in the caves occupied by people to see whether these might give some clues.

The region is at a junction between the mid-latitude belt that we identified in the previous chapter and the southward prong that leads to the southern tip of Africa. Here three main kinds of environments—Mediterranean woodland, dry steppe, and subtropical desert—are in close juxtaposition and the area covered by each fluctuated as temperatures and rainfall changed.[7] There are clear signals that the fortunes of the animals typical of each of these environments waxed and waned as they reacted to changing conditions.[8] The evidence has been unduly forced, in my opinion, to favour the idea that proto-Ancestors came into the region from the south as part of a northward dispersal of the African fauna.

Too much emphasis has been placed on the movements of faunas when it has become abundantly clear that the mammals of the Pleistocene responded to changing conditions individualistically and not as a fauna.[9] According to these interpretations, not only did proto-Ancestors arrive from the south with an African fauna but the Neanderthals arrived from the north with a Palaearctic fauna.[10] What can we really say, from the animals found alongside humans in the caves of the Middle East, about the climatic and environmental conditions at the time when these people were living there? The answer seems to lie in variations in local conditions, largely induced by rainfall and drought.

The Middle East is today an area of the world in which water is the most important ecological limiting factor.[11] Water seems to have been the critical factor during the last interglacial too, when Neanderthals and proto-Ancestors occupied the region. The best preserved record of animals associated with proto-Ancestors comes from Qafzeh,[12] so let us see what kinds of animals were around then. Among the most common herbivores were wild cattle and red deer. Neither species can be defined, by any stretch of the imagination, as African in origin. They were instead species typical of the middle temperate latitudes—they would have occupied a wide range of habitats, largely open woodland and shrublands. Next was the Persian fallow deer, which is also at home in light woodland, for example among Tamarisks. Its natural distribution was across north-east Africa to present-day Iran so we cannot consider it an indicator of African arrivals either.

The pattern repeats itself with other herbivorous mammals. After the top three we find the bezoar goat, the wild boar, and the mountain gazelle. These are all species of the mid-latitude belt: the goat is a species of rocky habitats from Asia Minor to the Middle East to Sind;[13] the gazelle of light woodland from Egypt, the Middle East, and the Arabian Peninsula to Iran; and the boar of open woodland and shrublands and was widespread across large areas of temperate Eurasia. These species talk of rocky habitats and open woodland still to be found in the area today. We could probably add the extinct narrow-nosed rhinoceros to this group. It is a species that was found in Qafzeh in a smaller proportion but which was nevertheless quite common and it too lived in savannah-like, open woodland, as well as across temperate Eurasia.

There were horses here too and the presence of an extinct North African horse has been used in support of the African connection.[14] This horse was, like the species that we have seen so far, instead a species of the mid-latitude belt. We know little about its habitat but it must have in all likelihood included open or semi-open grassland. Three species, scarce in the fauna of the cave, are nonetheless added to the African support list: the Hartebeest is a species that once lived in grasslands right across Africa, as far north as Morocco, so its presence in the Middle East is not surprising as this area would have been on the northern fringes of its natural range. Similarly, the hippopotamus was once widespread across Africa and would have occupied the northern extension of the Rift Valley in the Middle East so long as water was available. Finally the Dromedary is not African at all. Its natural range is uncertain, as all existing dromedaries are domesticated, but archaeological evidence suggests that it included the Arabian Desert at least.[15] The presence of a large number of ostrich egg shells in Qafzeh has also been cited in support of an African connection but ostriches were also once widespread across North Africa. There were ostriches in the Middle East right up to 1914 and surviving populations in the western Sahara live today in desert-steppe.[16]

Similar arguments have been advanced using the small mammals found in Qafzeh, for example the absence of hamsters and Palaearctic voles or the presence of gerbils among others,[17] but these only serve to show presence of particular habitats and absence of others. The combined picture that emerges from Qafzeh is of a mosaic of habitats, probably at a particular junction between the three main habitats of the Middle East. The main species indicate Mediterranean woodland but not dense forest,[18] smaller numbers of other species indicate dry steppe and subtropical desert, the goats show that there was rocky habitat in the area, and the hippos indicate the presence of standing water that may have been seasonal. The fauna of the other proto-Ancestor site from the last interglacial, Skhul, reveals a very similar picture.[19]

Were contemporary Neanderthals living in similar environments? We only have the site of Tabūn for comparison with Skhul and Qafzeh. The dated Tabūn Neanderthal lived around 122 thousand years ago.[20] The fauna is similar to that of Skhul and Qafzeh although the proportions of different species change. It is tempting to interpret the absence of arid steppe and desert species as reflecting a climatic difference, with Neanderthals occupying the region during a warm and wet period, but these species were not, in any case, abundant at Skhul and Qafzeh so their absence may simply reflect poor sampling of the rare species in Tabūn. What is striking is the dominance of Persian fallow deer and, to a lesser degree, mountain gazelle. This suggests that the Neanderthals of Tabūn were eating species that they were finding in open, savannah-like woodland. This interpretation receives support from the study of phytoliths, tough microscopic plant remains, which showed that the area around Tabūn Cave at the time of the Neanderthals would have been a Mediterranean-type woodland.[21]

Proto-Ancestors appear to have been less focused on the relatively small species that the Neanderthals were consuming and seem to have been taking, instead, larger animals in denser woodland as well as species out on the open steppe and semi-desert. With information from so few sites, any conclusions about behavioural differences among Neanderthals and proto-Ancestors must, however, remain tentative. One thing is for certain on present evidence: with the start of global cooling around 80 thousand years ago the proto-Ancestors disappeared from the scene altogether. Between then and 38 thousand years ago we have no secure evidence of Ancestors of any description in the Middle East despite the fact that this was the time of the expansion of the Ancestors from Africa (Chapter 4).[22] It is instead the Neanderthals that we continue to find in the region during this time.

Two sites are critical to our understanding of the post-80-thousand-year Middle Eastern Neanderthals. They are Kebara on Mount Carmel and Amud, north-west of the Sea of Galilee. The archaeological levels associated with the Neanderthals at Kebara span the time range between 75 and 40 thousand years ago, and those at Amud fall between 81 and 41 thousand years ago.[23] The dates from Kebara suggest repeated occupation of the site, mostly after 70 thousand years ago and up to 43 thousand years ago. These dates fall within a cooler period than the preceding interglacial but they encompass a relatively mild time marked at either end by severe cold and dry conditions, at 70 thousand and between 47 and 42 thousand years ago.[24] The majority of the Amud dates also fall within this time frame and two dates for levels with burials put these at 53 ± 8 and 61 ± 9 thousand years ago, respectively. Taken together, these dates suggest that the Neanderthals were in the area during a period cooler than the preceding one but that they may not have been about during the coldest and most arid moments.

There are more sites that were occupied in the Middle East during the last interglacial and the later period of climate cooling but most are known only from stone tools and animal remains. The problem is that the Neanderthals and the proto-Ancestors both made similar tools,[25] so that, in the absence of fossils, we cannot discern the makers in these sites. In fact, similar tools are found across North Africa and right down to South Africa,[26] and they indicate a technological change from the globally widespread hand axe industries that preceded them and which were made by *Homo erectus* and, later on, *H. heidelbergensis*.

No Neanderthal remains have been found south of the Middle East and it is therefore assumed that the entire continent of Africa was occupied by a single species that I have defined as proto-Ancestor. Identifying humans is not an easy task though. We have already seen the difficulty

in defining specimens that appear ancestral but retain archaic characteristics and this is true of all Ancestors that we find prior to 38 thousand
years ago.[27] One way of dealing with this obvious difficulty, imposed by
the pigeon-holing philosophy of classification that cannot adequately
deal with the continuous way in which life forms evolve, has been to
separate anatomical and behavioural modernity. In other words, people
started to become anatomically modern (our proto-Ancestors) but they
were not really fully modern (our Ancestors) until they achieved the
behaviours that we would recognize in ourselves.

The attempt to separate anatomy and behaviour has been a response
to the increasing fuzziness among the fossils that we can place as fully
ancestral against those that are on the way to becoming Ancestors but
still retain features that betray an ancient link with one of the populations of *H. heidelbergensis*. The neat evolutionary story that led to the
idea of a human revolution no longer stands up against the evidence.[28]
The changes from proto-Ancestor to Ancestor were, instead, gradual,
varied between regions, and took place over a long period of time. The
goal posts have shifted position and the debate has moved into one
concerning when we first detect evidence of ancestral behaviour in the
archaeological record.

One school of thought in this new debate proposes a sudden arrival
of Ancestors, which is placed at 50 thousand years ago; they argue
that 'it is surely reasonable to propose that the shift to a fully modern
behavioural mode and the geographic expansion of modern humans
were also co-products of a selectively advantageous genetic mutation.'[29]
I have difficulty with this idea for the simple reason that there is
absolutely no evidence of any mutation that suddenly made our ancestors modern. The gradual emergence of behaviours that we identify
with the Ancestors seems a more plausible alternative.

As so often happens in these debates, the argument has become
polarized and the opponents of the sudden emergence of modern
behaviours seem to have spent a disproportionate effort on a pilgrimage
of discovery trying to demonstrate the opposite. The problem starts
with defining what modern behaviour is, and it does not mean the same

to all those trying to find it.[30] Beads have recently become the subject of attention as one of the hallmarks of modern behaviour. Together with art, personal ornaments are considered 'unquestioned expressions of symbolism that equate with modern human behaviour'.[31] So we turn to beads.

A number of recent discoveries claim to show evidence of modern human behaviour, defined by the presence of beads or art, stretching back to 164 thousand years ago in South Africa.[32] The finds are not limited to South Africa. Perforated shells of marine molluscs, claimed to have been worked by human hand and used in necklaces (no necklace has actually been found), also come from North Africa and the Middle East and they are all supposedly older than 73.4 thousand years ago.[33] They are all well before the, yet to be found, 50-thousand-year-old mutation that apparently made our ancestors fully modern.

To summarize, we find beads in archaeological sites in North Africa, the Middle East, and South Africa and they span the huge period from roughly 130 to 73 thousand years ago. Additionally, worked pieces of ochre have been found in South Africa and they go back even further, to around 164 thousand years ago. Can we be comfortable with a definition of modern human behaviour based on perforated shells and bits of worked ochre? It becomes especially worrying when we look at the discoveries in greater detail.

For South Africa, what we have are 39 perforated shells of a mollusc (*Nassarius kraussianus*) from a Middle Stone Age level at Blombos Cave dated to 75.6 ± 3.4 thousand years ago.[34] The human remains from this site consist of a number of dental fragments insufficient to confirm with certainty that they belonged to proto-Ancestors.[35] Next we have two perforated shells of a related species (*Nassarius gibbosulus*) recovered from the old collections at the Natural History Museum in London.[36] These were from Skhul which we have seen was a site linked to proto-Ancestors. Elaborate tests were made of some sediment remaining attached to the shells; this was linked to sediment samples kept from the early twentieth-century excavations at the site, from where the human remains had been found, because there was no longer any material that could be excavated back at this site. The potential errors inherent in dating a sample from old collections seem obvious; yet this is the way in which two little shells from Skhul, with holes on their sides, have been used to prove that proto-Ancestors in the Middle East were behaviourally modern.

The North African evidence is not overwhelming either. There is a single perforated shell of *N. gibbosulus* (the same species as the two from Skhul and also from an old collection, in the Musée de l'Homme in Paris) from Djebbana in Algeria, a site for which we have no convincing evidence of its date and which has no associated human remains.[37] Finally there are 13 perforated shells, also of *N. gibbosulus*, from Taforalt in Morocco.[38] These have been claimed to be 82 thousand years old but the age range for the archaeological levels in which the shells were recovered spans from 73.4 to 91.5 thousand years ago so they could be from anywhere in this nearly 20-thousand-year-old time span. It is also assumed that the makers must have been proto-Ancestors, only on the basis that no Neanderthal fossils have been found in North Africa. The remains of the makers of the Taforalt beads have simply not been found.

In sum, 55 perforated shells from one South African, one Middle Eastern, and two North African sites, that span 57 thousand years of human history, is what we have as proof that people were behaviourally modern prior to 50 thousand years ago. Because we have found none in Neanderthal sites of comparable age we conclude that the Neanderthals were behaviourally archaic, in other words not very bright. This kind of evidence is, as we shall see later in this book, the best that can be offered to demonstrate how cognitively superior, behaviourally modern, Ancestors caused the extinction of the primitive Neanderthals.

It has also been argued that these shells show that the practice of perforating them for necklaces was widespread among proto-Ancestors across Africa at the time. Pinnacle Point in South Africa, the site of the 164-thousand-year-old worked ochre, is also a site claimed to provide the first evidence of human exploitation of the coast.[39] If people first started to exploit the coast 164 thousand years ago, they could not have started making necklaces out of perforated marine mollusc shells before that time because they would have had no contact with them. That they seem to have started making the necklaces relatively soon after speaks not of cognitive abilities but rather of being in situations in which these shells were accessible to them.

We will return to the exploitation of the coast and its resources in the next chapter. I want to turn now, instead, to two major resources fundamental to the course of our evolution: grass and freshwater. The reason for doing so is because these were the two critical elements that facilitated the geographical expansion of populations of *H. heidelbergensis*, proto-Ancestors, and Ancestors. Grass and freshwater and not coastal resources, as is often suggested, were the main agents driving human geographical expansions for the greater part of our history.[40] Grasses were the trigger.

Plants today fix carbon dioxide during photosynthesis, the process by which they generate energy, in one of three ways. The two main ones are known as the C_3 and the C_4 pathways. The third method is used by fewer plants, especially cacti, and is of no great concern to us here. For much of the history of plants it was the C_3 pathway that was used and C_4 photosynthesis only appeared around 20–25 million years ago,[41] although C_4 plants remained scarce for a long time afterwards. The stars among the C_4 plants are the grasses and they do particularly well in warm climates and with low carbon dioxide concentrations in the atmosphere. C_3 plants, on the other hand, do better in cooler climates. Decreases in atmospheric carbon dioxide around 6–8 million years ago gave grasses an edge in warm environments and these plants started to spread in several regions of the world.[42] It was the beginning of the new world of grasslands and savannahs.

The appearance of C_4 grasses in Africa followed a pattern, starting 8 million years ago at the equator, where the warmest climates favoured by these plants were found, and reaching the cooler southern Africa 5 million years ago.[43] Even so it took a while before these grasses dominated entire regions. Significantly, in East Africa we observe an important shift in favour of C_4 grasses after 1.8 million years ago and the signal of open C_4-dominated grasslands is evident only after 1 million years ago.

This means that for the early period of proto-human evolution, between 8 and 1.8 million years ago large areas of Africa would have been a mosaic of vegetation which included a small element of grassland. It comes as no surprise that the early proto-humans lived in habitat mosaics, never far from woodland as we saw in Chapter 1. It may be no coincidence either that we first detect H. erectus precisely at the time, 1.8 million years ago, when the shift favouring C_4 grasslands in the landscape began. If we follow the logic of the argument put forward in the Prologue, then the advancing grasslands would have been the peripheral areas of the open woodland that was the core area of the small-brained proto-humans. It would be precisely in these edge areas that we would expect populations under ecological stress and where we would expect to find innovation.

Homo erectus took to the open grasslands but did not abandon the trees altogether. What we find is a move towards exploiting habitat mosaics that were mainly savannahs and open grasslands, which is where the large grazing animals were most abundant and easiest to locate and where a living could be made from a diverse diet that included meat. The denser woodland did not support as many animals and, in any case, these would have been harder to locate among the dense vegetation.

The premium territories would have been those that had these open habitat mosaics close to water. In these areas people would have not had to go far to drink and the water would have attracted potential prey. Wetlands are rich and productive environments and have been of vital importance in the evolution of life and have been widely associated with human activity.[44] It is no surprise that nearly all fossil sites that have been found associated with H. erectus, H. heidelbergensis, and proto-Ancestors have combined these features.[45]

By the time we reach H. heidelbergensis we find a powerful predator, capable of despatching large game by ambush hunting, well established on the scene. This powerful hunter then faced an evolutionary dilemma: become light and capable of ranging widely or keep a bulk that gave the edge in a world of close-quarter hunting of large grazers. Those that stuck to the tough build did well in habitat mosaics for a very long time but as climate deteriorated between 70 and 20 thousand years ago, many vegetated areas opened up and they all went extinct along with the large animals with which they had co-evolved. They included the Neanderthals in Europe and Asia, proto-Ancestors elsewhere, and *H. erectus* in Asia.

In *H. erectus* we had the beginnings of a body built for endurance running and walking, one that would become increasingly refined in some populations as time went by.[46] Among the populations that sacrificed power for a lightness that allowed wide-ranging behaviour were those that endured in the open, semi-arid lands of north-east Africa. This body has been seen as a design that improved the territory that could be covered in pursuit of prey or in reaching carcasses quickly. It is just as likely that in the dry and highly seasonal lands in which it became perfected, the ability to range widely offered other advantages.

Much of the evolution from *H. erectus* to proto-Ancestors and beyond took place in a water-limited world, one in which water was the dominant limiting factor on the productivity of the environment.[47] The ability to seek out ephemeral and widely scattered sources of water and to track the seasonal flushes of grass would have been high on the list of priorities. Once again evolution was active in the margins of the habitat mosaics that were the core territories of the tough *H. heidelbergensis* and some of their proto-Ancestor descendants. As increasingly large areas of this world came under the influence of rainfall–drought regimes and seasonal grasslands, humans that had survived in what had previously been marginal areas got lucky and thrived.

It is not surprising then to find proto-Ancestors in semi-arid open woodland, grassland, and steppe mosaics in the Middle East, in sites like Skhul and Qafzeh, between 130 and 100 thousand years ago. Their arrival there would have been part of a wider expansion of territory, probably from north-east Africa where the earliest proto-Ancestors have been found,[48] into adjacent regions of similar habitat. This expansion would have taken them very quickly across the semi-arid lands of North Africa, keeping to similar climates and habitats, all the way to the Atlantic coast of Morocco where they were established by around 160 thousand years ago.[49] This was an expansion across the western part of the mid-latitude belt but the sea and the mountains of the Middle East and the Neanderthals may have kept them away from the northern shore of the Mediterranean.

The early dates from Ethiopia and Morocco strongly suggest that an east–west expansion took place during cool and relatively dry conditions that would have favoured proto-Ancestors used to living in semi-arid environments.[50] It suggests that these early people must have been adept at surviving in such parched landscapes and had abilities that allowed them to range far across treeless landscapes in search of water and widely scattered food resources. This may well have been an early experiment at breaking away from the shackles of the open woodland habitat mosaics that was their heritage. It may have been the kind of rapid spread of a population that took place when favourable circumstances created optimal conditions and resembles expansions that we will see later in this book when we look at Australia and the Eurasian Plain.

Aridity within the African continent, the Middle East, and the Arabian Peninsula prior to 130 thousand years ago would have created a patchwork of habitats and would have restricted the rainforest to the central parts of the continent, as today. The resulting patchwork of woodland and savannah, broken up by semi-arid and desert lands, would have promoted the break-up of human populations into isolated nuclei. Some would have suffered and disappeared while others, like those that spread across North Africa, would have thrived or at least survived. We can find evidence of this early splitting of populations, between 190 and 130 thousand years ago, among living African people.[51]

Geneticists have looked at living populations from across the world to work out how we are all inter-related. These studies have become increasingly sophisticated and give us a good picture of our expansion across the world from a small ancestral population.[52] This global expansion will occupy us in later chapters. What is also very clear from these studies is that African populations are the most genetically diverse, which means that they have had the longest time to accumulate mutations. The presence of a particular mutation serves as a marker linking different living populations and the time of separation of the different populations can also be estimated (see Chapter 1). The overall conclusion that we can draw here is that all the genetic diversity accumulated prior to 80 thousand years ago is found only within living African populations. It is only after this that we find specific mutations not present among Africans and which show movement into other parts of the world.[53]

It is very possible that these pre-80-thousand-year-old geographical expansions within Africa affected adjacent areas. They also involved populations of proto-Ancestors as well as those of the earliest Ancestors. We have seen how proto-Ancestors reached the Middle East and it would not be surprising if they were also to be found in Arabia and India. But as these populations died out their signal was lost. The Skhul and Qafzeh people of the Middle East, with whom we started this chapter, show us that success was not always the outcome. These people vanished as the climate deteriorated after 100 thousand years ago. It may have been a loss of resources or of habitat, or they may have been ousted by other people, the Neanderthals. We may never know. Whichever way it was, these proto-Ancestors, beads and all, died out. It is probably just one out of many failed experiments that we may never fully comprehend. In the highly unstable world of chance and climate change many populations of humans simply vanished.

4

Stick to What You Know Best

I T is hard to imagine while strenuously walking in the heart of an equatorial rain forest, gasping for every breath in a stifling humid sauna, how people could have ever adapted to life under these conditions. It is not just the oppressive climate—the tall forest itself is dark, little light reaching the floor from the canopy, and you do not see any animals. It is a complete contrast to the herbivore-rich dry savannahs of tropical Africa. Yet there are many animals here, evident by the loud, continual noise of large cryptic insects and the constant threat of stepping on a deadly king cobra. This was my first impression of the rain forest in Borneo.

I was in Borneo on a UNESCO mission to look at a cave in the heart of the rain forest. Each morning I would wake up within a patch of relatively open vegetation, a clearing that housed the headquarters of the Niah National Park, 31 square kilometres of lowland forest situated 16 kilometres from the coast. Here amidst large, colourful butterflies and hornbills I got a reprieve from the claustrophobia of the prison of buttressed giants that made up the forest beyond. There is something in our nature that seems to crave for open landscapes, distant views of the horizon, and patchworks of trees and open spaces. Where we have not had access to these we have created them.

Gordon Orians, a biologist at the University of Washington in Seattle, has developed some ingenious tests in which he has shown children from different continents images of many different kinds of habitats. He has found that images of the African savannah are preferred above all others up to a certain age when preferences become coloured by personal experience of the place where the child has grown up. His 'savannah hypothesis' proposes that we have retained within our biological makeup a preference for habitats and landscapes in which we grew up as a species. It was in the savannahs that we found food and shelter for the greatest part of our evolutionary history and the past 10 thousand years of civilization have failed to stamp out the signal.[1]

Perhaps this was part of the reason why I felt uneasy in the deep forest. A 3-kilometre walk across the rain forest, which included a crossing of the opaquely orange-coloured Niah River on a flat-bottomed boat, was the daily trek to the Niah Great Cave. The distance seemed double that covered, due to the hot and humid atmosphere that I was unaccustomed to, but it did not seem to trouble the local Iban people who would walk past us, top heavy with sacks full of nests of the swiftlets that bred inside the cave. These nests were on their way to some up-market restaurant in a distant land where they would become the central ingredient of a fashionable delicacy—Chinese birds' nest soup.

The recent mass harvesting of the nests is causing a significant decline of the swiftlets. For many years the traditional method of taking the nests after the birds had raised their young, twice each year, had been sustainable but now there seems to be a free-for-all radically altering the balance. It is a cameo of human nature: traditional hunter-gatherer methods have been, by and large, sustainable. Once we found

Figure 8 Niah Great Cave in Sarawak, Borneo (Malaysia). This is the earliest known Ancestor site in south-east Asia and one of the earliest outside Africa

Photo credit: Clive Finlayson.

ways of dealing with surpluses, we seem to have inevitably chosen routes that rapidly exhausted our supplies.

Niah Great Cave is aptly named (Figure 8). Like many limestone caves in equatorial climates, Niah is huge. Over 3 kilometres of passages have been explored and the cave is, in places, 60 metres high and 90 metres wide. No light penetrates deep inside and it is impossible to move without torchlight. So why is Niah relevant to our story? It can hardly be because of its size or the swiftlets. In the previous chapter we saw how a young and energetic Dorothy Garrod had managed to make giant inroads into our knowledge of prehistory, particularly in the Middle East. It took another larger-than-life character to bring Niah to the forefront of prehistory.

If Dorothy Garrod's name is forever embossed in the history of the archaeology of Mount Carmel, then it is Tom Harrisson's that is engraved in Niah. Harrisson was born in Argentina in 1911 and was educated in Harrow and Cambridge as an ecologist. He pursued a kaleidoscopic range of interests from ornithology to exploring, journalism and broadcasting, film making and anthropology. During the Second World War Harrisson was recruited by the British Army and parachuted into Borneo with the instructions to use the native peoples of the forest against the Japanese. It was after the war that Harrisson stayed in Borneo and became curator of the Sarawak Museum.

It was during his time in the museum that he carried out pioneering excavations, with his wife, Barbara, in the West Mouth of the Niah Great Cave, between 1954 and 1967.[2] Among the most startling finds was the skull of a human (the Deep Skull) that was radiocarbon-dated to 40 thousand years ago.[3] At the time the results, representing the oldest Ancestors in south-east Asia, were viewed with scepticism by the archaeological community. Decades later, a team led by Cambridge archaeology professor Graeme Barker published results in 2007 that vindicated Harrisson's conclusions.[4] The humans of Niah Cave remain the oldest known Ancestors in south-east Asia. The success in Niah is a tribute to Harrisson's insight, perseverance, and endurance, given the particularly unpleasant excavation conditions in what came to be known as the Hell Trench.

The latest work in Niah has pinpointed the Deep Skull to a period between 41 and 34 thousand years ago.[5] It seems that people were living in the cave even earlier than this, at least as early as 46 thousand years ago, a time when Neanderthals were the only people living in Europe. All the various kinds of humans that we have met so far had been living in habitat mosaics that were relatively open, usually with some trees, very much in keeping with Orians' savannah hypothesis. Does this mean that the people in Niah were exploiting the rainforest, and was this a new development in our evolution?

The picture emerging from Niah is that people were exploiting a range of habitats around the cave, some of which are not to be found in the area today.[6] There was some rain forest in the area but for much of the time that humans were there, with the climate cooler and drier than today, most of it was replaced by dry woodland and savannah. It is difficult to imagine such a landscape where rain forest dominates today but the impact of the Ice Ages was felt even this close to the equator.[7] The landscape changed with the climate and it seems that forest returned on at least two occasions when people were living in Niah between 46 and 34 thousand years ago.[8]

It seems that the Niah people were at home in the open woodland and savannah but they were also capable of exploiting the rain forest. They were probably most at home in a patchwork of habitats that included some rain forest, a strategy that we already saw in the proto-Ancestors of the Middle East around 70 thousand years earlier (Chapter 3). It would have been in such mosaics that they learnt to exploit the rain forest without having to fully abandon life in the savannah. These people were sticking to what they knew best while learning new tricks. Let us have a look at what these new tricks were.

The Niah people were tailoring their hunting and gathering technologies to suit the habitats that they were exploiting. During warm and wet periods of forest encroachment into the savannah these people became more dependent on the forest. It would have been the same process of innovation on the periphery that we have seen already drive much of human evolution but, paradoxically, this time it was in reverse: people had by now become specialists of open woodland and savannahs and the deep forest was an inhospitable place, in the margins of their comfort zone. It was in the forest that these people would have been most challenged by their surroundings.

The overall food procurement strategy of these people seems to have focused on the mammals most abundant and these were wild pigs that would have roamed in the savannahs and also in the forest. The Niah people were continuing a long-standing tradition of the human lineage which was to catch medium-sized mammals for food,[9] but they took it a step further. Here monkeys and other primates were plentiful and they were hunted on a regular basis: the Niah people ate primates, including the large orang-utan.[10] It was a way of dealing with a new situation by venturing from the savannah into the forest edges. But we should not fall into the trap of seeing this as some kind of revolution. These people were simply doing what was characteristic of their species by now—they were innovators improvising in a new world.

The methods of catching these animals appear to have been non-selective. There is no evidence that animals of particular ages were singled out to be caught. Given that animals are hard to see and rarely cluster in herds or large groups in forested environments it seems very likely that trapping and snaring techniques had been developed by this stage. If so, this would explain why the prey animals taken in Niah did not fall into particular species or age categories.

The Niah people were omnivorous; they used much of what was available in the mosaic of forest and savannah that surrounded them. They took turtles and even monitor lizards. Large numbers of fresh-water fish and molluscs, taken from local rivers and swamps, were eaten but there is no indication that marine foods were consumed. During much of this time sea levels were lower than today and Borneo, Sumatra, and Java were joined to mainland south-east Asia, creating the land mass of Sundaland, which was roughly the size of Europe. Niah Great Cave would have been much further from the coast than now and it seems that the Niah people did not exploit it at all. There is a chance, of course, that they made seasonal trips to the coast and consumed marine resources there. After all, why would they transport fish and molluscs over large distances back to the cave? It is a possibility that, like so much of the fragmented jigsaw of prehistory, remains in the air.

The forest was also a resource of plants but many potential food sources were toxic and needed to be processed before being eaten. The Niah people were adept at dealing with these problems imposed by the chemical defences of the forest plants and they had the knowledge and technology that could neutralize otherwise highly nutritious forest plants, for example burying nuts in ash pits to detoxify them.[11] All this new evidence from Niah paints a picture of people who first arrived at the site some time after 50 thousand years ago, probably following the spread of savannahs and its animals from the north at a time when they could have walked from present-day mainland Malaysia to Borneo. Once there they managed to stay in the area and found ways of dealing with habitat alterations forced upon them by climate change.

In Niah we find novel ways of exploiting resources: catching primates that lived up in the trees, trapping and snaring, fishing and detoxifying plants. They are all examples of solutions developed under stress at times when the savannahs were being replaced by rain forest. The most startling of these novelties was the practice of burning down the rain forest itself. Large quantities of pollen from plants known to be the first to colonize areas ravaged by forest fires were found in Niah and these coincided with periods when the rain forest was expanding. Marine cores taken off the coast of Borneo also record a period starting around 50 thousand years ago, the time of the first known occupation of Niah, of abnormally high concentrations of microscopic charcoal particles.[12] The signals are too large and sudden to be solely the result of natural fires; it really does look as though the practice of burning the forest to create open savannah that was attractive to the animals these people hunted started this far back.

We are left in no doubt that from at least 50 thousand years ago, people that we can identify with—Ancestors—lived on the edge of the Bornean rain forest. Their behaviour shows flexibility and ingenuity of a kind that mirrors the way we are but they did not paint in the caves and they were not into beads either. Similar lack of evidence in contemporary European Neanderthal sites has condemned them as backward and archaic humans. Such is the subjectivity with which we have approached our evolutionary history.

In the previous chapter we looked at the pre-80-thousand-year expansions of proto-Ancestor populations within Africa and adjacent areas such as the Middle East. But how did people get from Africa 80 thousand years ago to Borneo 50 thousand years ago? When we look at living populations of humans from across the world we find that non-African populations all arose after 80 thousand years ago from a founder population that lived in north-east Africa, most probably Ethiopia.[13] It is here, then, that we must start.

It is interesting that this period of Ancestor expansion after 80 thousand years ago paradoxically coincides with the time when the Middle East appears to have been abandoned and occupied only by the Neanderthals. We have seen how climate cooled somewhat after 80 thousand years ago but remained relatively mild except for two periods of severe cold and aridity, at 70 and between 47 and 42 thousand years ago. At the same time the very wet periods that affected the eastern Sahara, the Negev Desert, and Arabia prior to 80 thousand years ago also seem to have subsided after this time.[14] This means that the period of geographical expansion from north-east Africa to south-east Asia was not particularly exceptional in terms of climate although it was becoming drier than earlier. It was neither warm nor wet, and severe cold and dry periods were limited. Let us keep these observations in mind as we try to understand the events as they unfolded.

When a new idea is first put forward it invariably meets resistance but sometimes, if it gathers advocates, it tends to creep into the accepted part of the science and often becomes dogma. This seems to be what has happened with the idea of the coasting movement of humans out from Africa. The idea of a 'southern route' out via the Horn of Africa into Arabia was first put forward by Marta Lahr and Rob Foley in Cambridge as far back as 1994.[15] It introduced the very novel and interesting idea that not all movements of people from Africa need have gone past the Middle East. The southern route has gained support in recent years; especially as genetic evidence has given us a clearer picture of how people dispersed.[16] The difficulty with the southern route started for me not with the route itself which seems sound but with the proposed way in which people used the route.

In 2000 a paper was published in the journal *Nature* that reported the interesting discovery of stone tools in an emerged reef terrace on the Red Sea coast of Eritrea.[17] The reef was dated to around 125 thousand years ago and the tools were apparently connected with the exploitation of oysters and other marine molluscs by proto-Ancestors. Here we had evidence of humans on the right stretch of coast, close to where Africa almost touched Arabia, at a time when they may have been about to launch themselves out of Africa.

At the time the discovery was highlighted as ground-breaking, providing the earliest evidence of human exploitation of the coast. In the previous chapter we saw how, by 2007, the time of the earliest exploitation of the coast by humans had been taken back to 164 thousand years ago.[18] Why collecting sea shells from the beach for food (and decoration) should strike anyone as extraordinary is more than a little odd as nothing particularly special or difficult is needed to perform such a simple task, as any beachcomber can bear out. Even the team that discovered the Red Sea tools on the reef seem to have realized later that there was evidence of human occupation of rivers, lakes, estuaries, and coasts close by which probably went back to the time of *Homo erectus* up to a million years ago.[19]

The reason why evidence of collecting food from the beach prior to around 125 thousand years ago is so difficult to find is that this was the time when sea levels were rising worldwide because of global warming. The rate of rise was rapid, reaching up to 2.5 metres per century, and the sea level went up to as much as 9 metres above present sea level.[20] It has not been this high since. It seems quite obvious that any coastal sites that had been used by people prior to this time were submerged and any artefacts were, in all probability, washed away. The cave at Pinnacle Point in South Africa, the 164-thousand-year-old site, survived intact because it was in a particularly steep part of the coast and high enough for the sea not to have reached it. It is the exception that proves the rule.

The most compelling evidence clearly showing that there is nothing very special about the gathering of food from the shore comes from observations made on the coast of Burma in the late nineteenth century. Alfred Carpenter from the Marine Survey Office in Bombay wrote to the journal *Nature*, where his observations were published on 19 May 1887.[21] In his note, Carpenter described how macaques that lived on the islands of the Mergui Archipelago in South Burma regularly ate oysters which they collected from the shore at low tide. These monkeys had developed the habit of striking the base of the upper valves of the oysters with a rock until they dislocated and broke up. The soft parts of the oysters were then picked out between finger and thumb. Stones that were to be used as hammers were selected and carried for distances of up to 80 yards, showing that this was not a random or unplanned affair. These observations seem to have been overlooked until 2007 when startling new evidence was published.

Thai scientists started a research programme along the Andaman Sea coast of southern Thailand early in 2005, monitoring the impact of the disastrous tsunami of December 2004.[22] Quite by accident they found two adult female long-tailed macaques using objects to crack mollusc shells on one island; on landing they found cracked oysters on the beach along with axe-shaped stones that the monkeys had been using. They began to wonder how widespread this astonishing behaviour was among the troops of macaques that lived along this stretch of coast. Further visits revealed that the monkeys regularly came to the beach to crack oyster and other shells with stones, just as Carpenter had observed 120 years earlier along the coast to the north of them.

The more they looked, the more familiar they became with this conduct. Macaques seemed to be particularly fond of crabs and they even observed a male peering into the water looking for them. The stone tools were used to crack oysters but also to detach other kinds of molluscs from the rocks. Overall, these animals were clearly omnivorous as they also ate figs, other fruits, and leaves. Talking to local people revealed that the behaviour of the macaques was a year-round affair and that, when they foraged for oysters in the mangroves where there were no suitable stones, the monkeys used the empty oyster shells to prise other oysters open. Had these been humans 120 thousand years ago, how much evidence would have been left behind for us to find?

The macaque lineage separated from that which would lead to humans some time between 21 and 25 million years ago,[23] an observation which would seem to put the notion of the proto-Ancestor discovery of coastal foods 164 thousand years ago on the limit of the absurd. If we then reflect that the stone tools made by humans on Flores, Indonesia—an island that has had no known land connection with the mainland—are older than 800 thousand years,[24] we can only conclude that people have been visiting the coast for a very long time and that there is nothing particularly special about their presence there as recently as 164–125 thousand years ago.

The long-tailed macaque, primate beachcomber par excellence, can teach us another lesson. These monkeys have managed to establish viable populations on a number of remote islands over a wide area of south-east Asia.[25] They even reached the Nicobar (south of the Andamans) and Philippine Islands, which were never connected to the mainland. The exceptional seafaring skills of these macaques seem to be linked to their habit of living mainly along riverside and coastal forests including, as we have seen, among mangroves. Natural rafts start their life in many of the large rivers of south-east Asia and they drift seawards where they are moved between islands by currents. The long-tailed macaques seem well equipped to disperse successfully over these water barriers and probably did so unintentionally on a number of occasions by getting caught in these natural rafts. In this case over-water dispersals to offshore islands were chance events that recurred in time.

Nobody, to my knowledge, has suggested that these macaques had found ways of making canoes or other watercraft and they do not seem to have developed maritime navigation skills either. The simple combination of their habits, which often brought them close to drifting rafts, and chance allowed them to populate many distant islands. Yet when it comes to the dispersal of humans across these same islands and onto Australia the prerequisites in all accounts of the epic journeys are watercraft and navigation skills.[26]

The aftermath of the 2004 Asian tsunami produced many incredible stories of survival, including people drifting out to sea. In one case, a 21-year-old man survived at sea for two weeks, eating only the flesh of old coconuts while on a raft of floating debris. This case serves to show that chance human movement across water barriers may have been, as with the macaques, quite feasible. If people, like the macaques, had been living along coastal areas and rivers in south-east Asia then they would have been prone to drifting out to sea on natural rafts too. Niah Cave has shown us already that not only were people living close to rivers over 40 thousand years ago, but that they were regularly collecting molluscs and catching fish from these aquatic environments.

What all this shows to me is that, whether or not people exited north-east Africa along the coast into Arabia and from there India and south-east Asia, it was not the result of a novelty. Humans did not suddenly discover the coast and its foods and they had certainly not been prevented from leaving Africa until they found this way. If it was not because of a new property that made these people modern and capable of conquering the world, then why did it happen when it did? To find the answer we need to look at conditions in north-east Africa after 80 thousand years ago, the time that seems to mark the beginning of mutations present among non-African people today.

It will be useful to recall for one moment the way in which animal populations spread from a source to nearby areas, gradually expanding geographically if conditions suit them. In the Prologue I used the example of the spread of the collared dove across Europe, from east to west, and I explained how this was an expansion that involved many generations of the birds. It was not a migration. Something similar would have happened with the extension of the geographical area of the people of north-east Africa 80 or so thousand years ago. It was not a migration, never purposeful, and certainly not an exodus in search of greener pastures. I stress this because the prevalent manner in which the expansion of human populations from north-east Africa in the direction of Australia is still portrayed as being an epic migration of peoples. The search for 'the route' is almost as pointless as the nineteenth-century quest for the missing link.

The first hurdle of the epic is the Gate of Tears, Bab-el-Mandeb, the narrow neck of water that separates north-east Africa from Arabia. This 25-kilometre-wide, 137-metre-deep, stretch of water where the Red Sea opens into the Indian Ocean has never been a land bridge. Even at times of lowered sea levels there would have been a narrow, 5-kilometre-wide, channel of water separating Africa from Arabia.[27] Was this narrow channel a barrier to people? It seems unlikely that it would have been but, in the absence of direct evidence, the question must stay open. If they did not get into Arabia across the Gate of Tears then they must have spread northwards into the Sinai Peninsula and from there across Arabia, or back down along the opposite coast.

Let us pause for a moment to consider how long a geographical expansion from north-east Africa to Arabia would have taken if the sea had been a barrier. To get from one side of the Gate of Tears to the other the long way round is a distance of around 4.5 thousand kilometres. At the rate of population expansion that we calculated for humans in the Prologue, which is of the same order as that calculated in other recent studies,[28] it would have taken people 1500 years or 100 generations. Getting into the Arabian Peninsula, as opposed to the other side of the Gate of Tears, would have required much less time; the distance of around 2.6 thousand kilometres would have taken approximately 876 years, or 58 generations.

So getting from north-east Africa into Arabia (or back for that matter) by a normal process of geographical expansion, assuming conditions were favourable for population growth, would not have required a feat achievable only by intelligent humans. Earlier versions of humans got all the way to south-east Asia much sooner, as we saw in Chapter 2, and many other species of animals have frequently managed to spread across even larger distances when conditions suited them. It would also be wrong to assume, as is frequently done, that the geographical expansion into Arabia was a movement in a single direction. It is far more likely that, from a core area, the expanding population of humans kept moving into suitable habitats in whatever direction these were. When looking for evidence of the post-80-thousand-year geographical spread of humans we should look in all directions, starting with areas closest to the core. The Nile Valley seems a good place to begin.

As we will also find when we move to Arabia and India, human fossils from this period are practically non-existent in this part of the world. This means that we must rely on stone tools and other evidence of material culture that testifies to the presence of humans. For the period that interests us here, the technology found in the region is part of the Middle Palaeolithic group, similar to the tools that Neanderthals and proto-Ancestors were making in the Middle East between 130 and 100 thousand years ago. In contrast to the Middle East, Middle and Lower Egypt appear depopulated at this time; at least, archaeological sites prior to 100 thousand years ago are virtually absent.[29] This seems strange as this is a time when climate was warm and wet across the region.[30]

Human presence in the Nile Valley becomes obvious after 100 thousand years ago.[31] At this time the Middle East was occupied only by Neanderthals so there was a kind of north–south divide between them and the Nile Valley people. Between then and 70 thousand years ago two different technological groups lived in the valley and these may have represented different people.[32] One group, the Nubian, was widely distributed across northern Egypt and they appear to have been at home in the desert as well. Their technology included stone points that would have been used to hunt game, apparently their main subsistence strategy. These Nubians made a sudden appearance in the northern reaches of the Nile Valley and it seems that they may have been outsiders who entered the area from the south.

The second population is the Lower Nile Valley group, apparently with local technological roots, and appears to have persisted alongside the newcomers, only rarely venturing away from the river. It may be that they were part of the proto-Ancestor population that had lived close by in the Middle East and had reached west as far as Morocco (Chapter 3); their technology was virtually identical to that of these proto-Ancestors, and the Neanderthals for that matter, which might lend some support to this view. Could the Nubians have been Ancestors? In any case the Nubian newcomers did not have an immediate advantage over the locals and both groups lived in the same region, doing different things in different habitats, for a long time. A third population, the Neanderthals, were not far away, to the north, either.

Occasionally, when climate brought their favourite environments together, the two Nile groups lived in close proximity. The conclusion that we can draw from this interesting study is that the dynamics of the human populations at the time were not that different from today: there was a dense population in the river floodplain and there were mobile, nomadic, groups in the adjacent deserts. The desert Nubians were flexible in terms of the way that they lived, moving around in times of stress and even becoming temporarily sedentary close to the river or the Red Sea coast. The sedentary river people kept well away from the desert and came into conflict with the desert people when they crossed paths.[33] In terms of the way in which I defined the actors of the human story in the Prologue, the Nubians were certainly the innovators and the others were the conservatives. In the constantly changing world of the Nile Valley, eastern Sahara Desert, and Red Sea coast, neither could gain supreme advantage over the other.

The archaeology of the Arabian Peninsula for the critical period of interest to us is unfortunately not very well known. There are sites with stone tool industries that resemble those of the Middle East, Africa, and India and they cover a wide time frame, from 175 to 70 thousand years ago.[34] Some of these sites are on the Red Sea coast but there are also others well inland, even in the mountains. The human occupation of Arabia at this time is closely linked to freshwater sources: rivers, streams, lakes, and springs but as climate became more arid people had to adapt to life in the desert. There is little by way of convincing archaeological support for a coastal movement of people towards India. If anything, the available evidence shows a widespread population, at least during times when grasslands and wetlands were common. The Arabian Peninsula was simply the eastern end of the semi-arid, seasonal, savannahs of northern Africa and its people must have been very closely related.

The presence of a North African technology, the Aterian, at a site on the south-western edge of the vast Rub' al-Khali (the Empty Quarter)

Desert illustrates clearly this connection.[35] This desert covers much of the south-eastern Arabian Peninsula and is a considerable distance away from the North African stronghold of the makers of the Aterian. A defining characteristic of the Aterian is a leaf-shaped stone point with a stem at its base where it would have been attached to a wooden shaft. The people who invented the Aterian were hunters and would have launched these projectiles at their prey from some distance. The technique represented a development from the stone points that the proto-Ancestors and the Neanderthals of the Middle East made and which were hafted onto wooden shafts to make thrusting spears.[36] The difference lay in the environment in which people were hunting and it does not mean that one group of people was brighter than any other. As this environment became drier and more open, with fewer trees and bushes from where to launch a close-quarter ambush, heavy thrusting spears were superseded by lighter ones that could be hurled from some distance. We will see a parallel example later on the Eurasian steppe.

Aterian sites first show up in the archaeological record after 85 thousand years ago and they became widespread across the Sahara, west to Morocco, and south to Lake Chad and Niger.[37] Wherever they have been found they seem invariably connected with deserts, even in Morocco where climate was semi-arid to arid, and the technology has been interpreted as an adaptation to the desert environment. Perhaps, more than a desert adaptation it was a way of hunting in arid, treeless habitats where the commonest herbivores were small- to medium-sized gazelles that could only be caught by having either spears or arrows launched at them from a distance.[38] As climate became increasingly arid in central areas of the Sahara Desert, the Aterians could not survive there any longer and died out; the later Aterian sites appear in places like Morocco which, though arid, retained some vegetation and fauna. Here, the Aterian may have survived as recently as 20 thousand years ago but this claim remains controversial.

Put together, the Aterian technology seems to have arisen just as the climate across the Sahara, north-east Africa, and the Arabian Peninsula was becoming drier. It is a technology that appears to have developed from the technologies that proto-Ancestors and Neanderthals had been making for thousands of years and it is probable, though the evidence is limited, that it was the proto-Ancestors that invented it.[39] In the Nile Valley, the practice of making leaf-shaped stemmed points seems to have started among the Nubians who were dispersing into large territories, into arid lands, at a time when climate was deteriorating.[40] It was a time when some proto-Ancestors, with traditional technologies, had abandoned areas like the Middle East; the habitat mosaics that had formed the backbone of our evolution were disappearing as steppe and desert engulfed the land. Many of the traditional, conservative, humans remained locked in islands of habitat mosaics close to the Nile River and shrinking lakes. A new world of arid, open spaces beckoned innovative humans that had survived in the arid periphery of the lush savannahs and woodlands of north-east Africa. Like the poor people of nineteenth-century Gibraltar whom we met in the Prologue, these innovators, born and accustomed to surviving in arid lands, coped much better with the stresses of a drying world than those who had had it very good for very long. Could the Nubians-Aterians have been making the transition from proto-Ancestor to Ancestor in North Africa and Arabia?

The usual interpretation of the exit of Ancestors from north-east Africa has involved expansion into lush savannahs or movement along coasts. It seems far more likely that the initial expansion was of people with flexible behaviour and strategies that made life in semi-desert and steppe possible. With climate change favouring aridity, the geographical area of these people enlarged from their north-east Africa stronghold. This expansion seems to have followed a wide latitudinal band across arid lands, west to the Atlantic coast of Africa, and east to the Indian Ocean coast of Arabia. The first step of the global expansion of humans was not an out-of-Africa movement towards Australia; it was instead, an out-of-Ethiopia spread across the arid lands of North Africa and Arabia, and possibly beyond into India. For now this spread of people could only have been at the expense of proto-Ancestor populations as the Neanderthals remained largely out of reach to the north.

The westward expansion of people would have been checked by the Atlantic Ocean, so Morocco marked the limit in that direction. Because the whole area between Ethiopia and Morocco is within Africa, this massive geographical spread of humans of around 5.5 thousand kilometres seems to have received little attention. Once again political boundaries seem to have conditioned our thinking. The equivalent spread eastwards would have taken people beyond Arabia, to the Ganges Delta on the doorstep of south-east Asia. It is this lap of the dispersal that has attracted the attention as the early out-of-Africa movement of Ancestors.

The Palaeolithic archaeology of India is more comprehensive than that of the Arabian Peninsula but there are few well-dated sites that can help us track down the arrival and expansion of proto-Ancestor and Ancestor populations.[41] Archaeological sites indicate that a substantial population existed in India prior to 100 thousand years ago, before the arrival of the Ancestors; these were presumably proto-*Homo sapiens*, perhaps related to those of Skhul and Qafzeh, but we cannot be certain of their identity. It is from traces in the genetic makeup of living Indian populations that we are able to pin down the arrival and subsequent movement of people.[42] The time of arrival of the Ancestors on the Indian subcontinent has been calculated at $64,828 \pm 15,000$ years ago. This is a broad estimate but is in keeping with a post-80-thousand expansion from north-east Africa. The growth of this population, producing local genetic variants within the subcontinent, appears to have been somewhat later, around $43,588 \pm 5,621$ years ago. It is perhaps no coincidence then that it is only after 50 thousand years ago that archaeological sites become, once again, more common and they seem to reflect a technological transition in the direction of the Upper Palaeolithic tool kits of the Ancestors.[43]

It is shortly after this resurgence of the Indian population that we pick up Ancestors in Niah and other sites in south-east Asia. What would cause an expanding Ancestor population's progress to be stalled on reaching India? The answer may be, once again, climate. We have seen clearly that the period around 70 thousand years ago was especially cold and arid globally. Even the specialized Aterians abandoned large areas of the Sahara at this time and it is likely that many areas of the Indian subcontinent became inhospitable. The growth of the expanding population would have been checked and scattered populations would have subsisted from a nomadic existence at low density, something that the archaeology would find difficulty in detecting. The demographic expansion that followed 50 thousand years ago may have been linked with a temporary improvement in the climate when savannahs stretched right across the subcontinent.[44]

It has been known for some time that the Ancestor population went through a genetic bottleneck,[45] in which populations must have dwindled, with later population growth after 50 thousand years ago. An intriguing suggestion has been that the bottleneck was caused by a volcanic eruption and the subsequent volcanic winter.[46] The volcano was Toba, on the island of Sumatra in south-east Asia, and it was the largest known explosive eruption the world has seen in the past 2 million years. The timing of the eruption, 73.5 ± 2 thousand years ago, seems to fit the bill. It left deposits of ash across southern Asia as far west as the Arabian Sea, in the Bay of Bengal, the South China Sea, and mainland India.[47] The volcanic winter caused a 'brief' 1000-year cold and dry episode that engulfed the region and this was followed by a longer, global, period of cold and dry conditions which was linked to glacial conditions. It was after this time, around 58 thousand years ago, that southern Asia became wetter once more with the return of the summer monsoons. For the people living in India, the humans closest to Toba at that time, the impact of the eruption, and subsequent ecological disaster would have been most severe. Followed as it was by glacial conditions, the period between 74 and 58 thousand years ago would have made life on the subcontinent difficult with only the hardiest people pulling through. It may come as no surprise that people used to living in harsh extremes may have been the ones that made it through the bottleneck.

For the people who made it through the bottleneck, whether volcanically induced or not, life on the Indian savannahs would have allowed rapid demographic growth which would, in turn, have triggered geographical expansion. We have seen how archaeological sites became much commoner in India after 50 thousand years ago, matching well the genetic evidence of growth and diversification. This growing population would not have found its limits within India. To the west they may have found the Thar Desert in north-west India a genuine barrier but to the south-east the rainforests of south-east Asia would have presented a new challenge.

Lack of water need not always be a bad thing. It depends where you are. Where there is a surplus of water, as there would have often been in the south-east Asian rain forests, dry periods would have opened up the forest and the savannahs would have encroached. Herbivorous mammals would have moved in and predators would have soon found them. The Ancestors would have been among the first to move in, rapidly exploiting the climate-driven windows of opportunity. During the coldest and driest moments a 50- to 150-kilometre-wide savannah corridor opened up on the, now submerged, Sundaland land mass.[48] It was probably by following the grazers across these savannahs that Ancestors reached Niah and the rest of the south-east Asian mainland.

The rain forest was never far away though and we have already seen how it returned to the vicinity of Niah at least twice, around 46 and 34 thousand years ago.[49] People would have now had to face new challenges of a kind never met between Africa and south-east Asia. We have seen how people had begun to adapt to the tough forest environment around Niah but we have also seen how they struggled to keep their preferred savannah homes by burning the forest down. In Niah we find a clear signal of resourceful people behaving like we would.

As the forest encroached Ancestors were sandwiched between their home-made clearings and the rivers and coast. In Niah they exploited the resources of the river and this may have been the real trigger to movement along watercourses, the natural roads within the dense forest, and the coast. Like the long-tailed macaques they risked being washed out to sea and in a similar sweepstakes game they probably inadvertently got across the many islands of south-east Asia.

The long-tailed macaques reached the Nicobar Islands and Ancestors reached these too and also the Andaman Islands; both are around 600 kilometres from the mainland and 250 kilometres from each other. It had been assumed that genetic evidence pointing to a pioneer human settlement of the Andaman Islands around 45 thousand years ago was evidence of a coastal dispersal from India to south-east Asia, part of the southern coastal route from Africa, but this evidence is now in doubt and the arrival may have been after 24 thousand years ago instead.[50] The colonization of the Andaman Islands by people, whether in purpose-built watercraft or on natural rafts, was simply part of a wider picture of Ancestor dispersal across the sea among a maze of islands. It would find its maximal expression, much later, with the colonization of Polynesia.

It is not difficult to imagine an unconscious island-hopping by people that would eventually get them into New Guinea. Once here, there was no water barrier preventing an expansion into Australia across what is today the Torres Strait. Aboriginal Australians, New Guineans, and Melanesians share genetic variations that are not found anywhere else, pointing to a single founder population of this region around 50 thousand years ago.[51] Once there, these people seem to have spent a substantial amount of time isolated from the rest of the world. To me this sounds like a population that colonized New Guinea from nearby islands by accidental rafting. If Australia had been reached instead across the 90-kilometre stretch of sea from east Timor using watercraft as is often suggested,[52] the question remains of why it was not repeated on other occasions or why Ancestors did not move in the other direction. Instead, once in Australia they remained trapped on an island continent.

The response to the arrival in this new land was the kind of demographic explosion that we have seen before, when conditions opened up new areas, and which we will see more of later in this book. Ancestors reached Lake Mungo, 2.5 thousand kilometres from the northern coast of Australia, some time between 50 and 46 thousand years ago.[53] The speed of expansion is a measure of how quickly populations can grow and expand under favourable and unlimited conditions. After 20 thousand years of post-Toba paralysis in India, people swept across south-east Asia, New Guinea, and Australia in virtually no time at all.

Once in Australia they may well have been trapped but they discovered a land of lakes, savannahs, and steppes, similar to the environments that generations of Ancestors had sought across North Africa, Arabia, India, and south-east Asia. They had stuck to what they knew best and found an unpopulated land. They brought with them innovations that had been picked up on the way—the use of fire to open up the landscape was one such skill that would be put to use on this new continent.

Meanwhile, in Eurasia, Neanderthals had also stuck to what they knew best—living in the mosaic landscapes of the mid-latitude belt. They too had managed to spread across these northern lands but they were trapped between the high mountains, deserts, and seas to the south and the cold treeless lands of the north. Luck had given the Ancestors a chance to reach places that the Neanderthals could not reach. In human evolution, success has often been the result of having been in the right place at the right time as we will see in the next chapter.

5

Being in the Right Place
at the Right Time

THE arrival of the Ancestors in Australia around 50 thousand years ago was the result of a coincidence of events and of people being in the right place, in this case India, at the right time. When a cooling and drying climate began to desiccate their savannah homeland, a new belt of savannah opened up to the south-east; the same climate change that was turning savannah into arid badlands in one place was reducing the domain of the rain forest and replacing it with lush grasslands, where tall trees no longer prevented light reaching the ground, in another. In came the grazers and predators, including people, soon followed.

We can speculate about the impact of these people on existing populations of *Homo erectus* that would have been around at the time of the arrival (Chapter 2). The last populations of *H. erectus*, if the dates are correct, may have survived in Java as recently as 25 thousand years ago which would suggest that the arrival of the Ancestors after 50 thousand years ago did not overwhelm them overnight. The many islands of south-east Asia may have acted as refuges for populations of archaic people and some seem to have survived until well after the arrival of the Ancestors. *Homo floresiensis* did so on Flores until 12 thousand years ago (Chapter 2). An even more remarkable discovery was reported early in 2008 from the island archipelago of Palau, north of New Guinea and east of the Philippine Islands. Around 25 skeletons of very small people were excavated and the stunning results revealed that they had lived as recently as somewhere between 2,890 and 940 years ago.[1]

These results reopened the debate on the status of the Flores people and posed the question whether dwarfism was indeed a common feature of humans that became isolated on remote islands. What is interesting for us here is that it shows that populations of humans of varying descriptions seem to have survived in isolation among the many small islands of south-east Asia, the world of hunter-gatherers and, later on, farmers having passed them by. It is just as plausible that populations of *H. erectus* survived in similar fashion until quite recently, having had no contact whatsoever with the newly arrived Ancestors.

In the previous chapter we saw how the Ancestors emerged from a population bottleneck after 50 thousand years ago. If the remaining populations of *H. erectus* in south-east Asia were affected by the same conditions that hit the Ancestors, be it climate change or the Toba eruption, then it is quite possible that many populations of these archaic survivors were pushed over the brink, leaving tiny isolated remnants that persisted in remote haunts. The Ancestors would have then entered largely empty space after 50 thousand years ago. The severity and speed of climate change would have been on a scale that tropical populations of *H. erectus* would not have been used to. If by then few of them were left, climate may have been the final nail in the coffin for most of them.

On the other hand the populations closest to Toba at the time of the eruption would have been the remaining ones of *H. erectus*; so this singular event would have hit them far worse than the Ancestors who were further away, in India. Perhaps the remnant populations of *H. erectus*, after a successful one and three-quarter million-year history, had found themselves in the wrong place at the wrong time. Curiously, the Neanderthals were far away from these lands and may have remained largely unaffected. The Ancestors would not have had it so easy in the north.

These examples illustrate emphatically what we have been finding throughout this book: that human history has been an affair between contingency and luck, conspiring with the erratic whims of climate and geology to produce the improbable character that is *Homo sapiens*. If we were to tell our story in the form of a Broadway musical, then the arrival of the Ancestors into south-east Asia and beyond would be the product of an improvised jam session; the extinction of *H. erectus* and the survival of *H. floresiensis* would follow a similar hit and miss plot.

History is typically the story of victors over vanquished and pre-history is no different. Because, out of the kaleidoscope of prehistoric humans, only we have made it to today, we have readily assumed the monopoly over the story; as survivors we seem to have chosen to portray ourselves in the role of victors and reduced the rest to the lower echelons of vanquished. To accept our existence as the product of chance requires a large dose of humility. Until now, we have instead preferred the self-centred course that highlights the supposed superiority of our immediate ancestors, like prehistoric conquistadores, over the rest. The Ancestors entering south-east Asia are assumed, in the absence of any evidence whatsoever, to have annihilated all other people except for the lucky minorities that hid in the jungles of remote islands. As we move north now into Siberia, Central Asia, and Europe, we will discover a much worse, and highly flawed, representation of prehistory along with a people who have been denigrated as the dumb brutes of the north—the Neanderthals.

There are many diversions in the study of human evolution that take us away from the central issues but the greatest red herring of all is the question 'Were the Neanderthals a different species from us?'[2] The question again stems from our obsession with pigeon-holing; it often distracts us and reduces the wonderfully fine-grained mosaic across space and time, the product of the permutations resulting from natural selection and contingency, to a few coarse pixels. As a result we lose the detail and misrepresent the processes.

At some point in our distant past we shared an ancestor with the Neanderthals. The development of technology in the 1990s that allowed the recovery of DNA from Neanderthal fossils opened up a window that enabled comparisons to be made with our own DNA.[3] By the end of 2006 the available technology and fossils had hugely expanded our knowledge of Neanderthal DNA and had opened up a new window that was expected to lead to the sequencing of the entire Neanderthal genome.[4] Already some headline-catching results have revealed the potential and have given us clues about the Neanderthals and the technique's potentials, which we could not have suspected when the first results were announced in 1997.

Among the sensational revelations have been those of hair and skin colour and speech. One study showed that the variability of Neanderthal hair coloration was similar to our own.[5] This included red hair in a proportion of the population, a result that inevitably generated 'Neanderthal red-head' headlines. The important point was that red hair was associated with pale skin. For Neanderthals living in Europe, such skin coloration may have been an advantage, allowing the production of vitamin D, which is mediated by ultraviolet light. The second disclosure was that Neanderthals shared with us two variants in the FOXP2 gene, a gene known to be involved in the development of speech and language.[6] The variants of the gene were thought to have been present in the common ancestor of Neanderthals and the Ancestors and seemed to point to the Neanderthals' capacities for spoken language.

A number of studies of Neanderthal fossil DNA have estimated the time of the last common ancestor with our lineage, when the two populations split up. The estimates are rough and have large errors. Some point to a split that took place after 600 thousand years ago.[7] It is hard to be more precise than this although a number of independent recent estimates seem to congregate around the 400-thousand-year mark.[8] Other estimates, however, put the split further back still, to 800 thousand years ago.[9] If the Pit of the Bones people (Chapter 1) were indeed distinguishable as a separate lineage from the Ancestors 500 thousand years ago and they were the ancestors of the Neanderthals, then the split must have predated these people, making an early date most likely. This means that, in broad evolutionary terms, the lineage that was to lead to the Neanderthals probably separated from our own soon after the onset of the 100-thousand-year climatic cycles that marked the start of the Middle Pleistocene 780 thousand years ago. This lineage probably emerged from one of the many populations that we currently amalgamate under the common name of *Homo heidelbergensis*.

So when Neanderthals and proto-Ancestors first met in the Middle East 130 thousand years ago (Chapter 3) they may have been genetically separated from each other, assuming no previous contact that has so far remained undetected, for over half-a-million years. The second contact, this time with the Ancestors themselves in Eurasia, some time after 45 thousand years ago, was even longer after the split. Whether these populations, on meeting, mated with each other remains a mystery: the pertinent questions to ask are whether there were enough people around for such contacts to have happened frequently; whether differences in the habitats and geographical regions that they occupied kept them apart; or whether biological and cultural differences between them prevented otherwise fruitful genetic exchanges. We shall look at these questions more closely in the next chapter.

Most texts on the Neanderthals consider the last interglacial, around 125 thousand years ago, to mark their heyday; it is the time when all the features of the 'classic' Neanderthals are clear for all to see.[10] The Neanderthals were an evolved population, descended from a lineage of Middle Pleistocene humans that had spread across a vast region of Eurasia, from Portugal in the west all the way at least to the Altaï Mountains of southern Siberia in the east.[11] These populations, part of the *H. heidelbergensis* group, lived in Eurasia during the Middle

Pleistocene, between 600 thousand years ago (or perhaps earlier) and around 200 thousand years ago,[12] and experienced a world very different from anything that had come before.

These large, tough, and intelligent people, whom we briefly met in the Pit of the Bones in Chapter 1, had evolved alongside a diverse fauna of large herbivores and carnivores at a time when the Ice Ages had started to have an impact on the world. This was a time when the Earth's climate became dominated by 100-thousand-year cycles that went from cold glacials to warm interglacials and back again. Overall, the world of the Middle Pleistocene was colder than before, especially after 400 thousand years ago, with brief interglacials that lasted, on average, about 10 thousand years.[13] Each glacial (Ice Age) ended rapidly with a sharp global warming towards the interglacial; the end of the interglacial was then followed by a gradual cooling towards the next glacial. Some of these interglacials were wet and heavily influenced by oceanic climate from the west while others came under strong continental control and were much drier.[14] Superimposed on this pattern were shorter pulses of cold and warm conditions.

The ecological changes that accompanied the onset of the 100-thousand-year climate cycles, around 780 thousand years ago, brought about a drastic reorganization of the species of mammals that lived in Eurasia. Many species went extinct, some being replaced by immigrants better suited to the new conditions, others evolving into new forms, and others leaving empty space. What followed were hundreds of thousands of years during which some species repeatedly expanded and contracted their geographical ranges. There was extinction and immigration while some species managed to evolve to suit the conditions of this new world.

These drastic climate changes generated barriers for the movement of species but also opened up land bridges as sea levels dropped during periods of lowered temperatures. Connections between different parts of the great Afro-Asian land mass became severed in critical places. Tropical areas became cut off from temperate regions to the north except in the easternmost parts of Asia, east of the Himalayas, where a connection between temperate and tropical climatic zones was maintained. Elsewhere, the high mountains, seas, and deserts that ran parallel to the latitude bands effectively cut areas off from each other.

The great mountain chains of the Himalayas, Hindu Kush, Pamirs, and Karakoram blocked off access to southern Asia from the north and west. Here *H. erectus* appears to have persisted in tropical areas of south-east Asia, largely isolated from others to the north and west and making repeated incursions north into temperate China when climate permitted. This connectivity between temperate and tropical areas of eastern Asia may account for the long persistence of *H. erectus* here.

Conditions in Africa were much drier than those in south-east Asia. The growth of the Saharan and Arabian deserts repeatedly sealed tropical and southern Africa from the north. Even when northward expansions were made possible by a wet and mild climate these populations found it difficult to penetrate northern Eurasia because of the barrier presented by the Mediterranean Sea and the tall mountain chains of western Asia (Taurus, Zagros, Caucasus). In tropical and southern Africa, populations of *H. erectus*, or descendant forms that we may collectively describe as African *H. heidelbergensis*,[15] became isolated. The climatic changes that affected the north were translated here into cycles of high rainfall and drought that created situations of local extinction, redistributed areas of occupation, and enabled local evolution to take place in response to new ecological opportunities.[16] The proto-Ancestors would emerge around 200 thousand years ago from this African melting pot.

Meanwhile across northern Eurasia, from Portugal in the west to Siberia in the east, populations of *H. erectus* were evolving into a new form, recognizable by his stature, robust build, and large brain. By 600 thousand years ago they are recognizable as distinct from *H. erectus*. These populations were evolving alongside animals that were new to the scene. The climatic cycles that started around 780 thousand years ago were relatively moderate compared with the rapid and severe changes to follow after 400 thousand years ago. Animals had time to adapt to changing conditions and we see a number of cases of gradual evolution at this time. The mammoths provide us with a good example.

Mammoths had dispersed from Africa into Eurasia and North America around 2.6 million years ago. The main species of mammoth that lived across northern Eurasia at this time was the southern elephant.[17] This was a species that lived in a temperate climate and inhabited wooded steppe. Between 1.2 and 0.8 million years ago, southern elephant populations living in north-eastern Siberia were experiencing the effects of climatic cooling and were adapting to living in herb- and grass-dominated habitats with permafrost. These mammoths were the first to be regularly exposed to the new climatic conditions and their teeth were changing to allow them to graze on tougher plants in new environments. These early Siberian mammoths were much larger than their ancestors and their fossil remains are recognizable as a new species, the steppe mammoth. It is some time after, with the spread of the new environments south and west as conditions became harsher, that the steppe mammoths entered Europe and replaced the southern elephants still living there as recently as 700 thousand years ago. The steppe mammoths became inhabitants of cool, dry, steppe with scattered trees, a landscape becoming increasingly widespread with the cooler and drier conditions.

The process was repeated once more, starting in north-eastern Siberia. The climatic deterioration that opened up new habitat for the steppe mammoths in the south and west made conditions harsher still in the north. By the time steppe mammoths were reaching Europe at the start of the Middle Pleistocene there were others in north-eastern Siberia that had evolved and were recognizable as a new species, the woolly mammoth. For several hundred thousand years steppe mammoths inhabited the south-west and woolly mammoths the north-east. Then something changed.

Around 200 thousand years ago the climate had worsened sufficiently to cause the westward geographical expansion of the steppe-tundra habitat of the woolly mammoth. But there was still good habitat for the steppe mammoth in Europe too so we find that both species lived alongside each other for a while. This was a relatively short-lived affair, however, and all the European mammoths after 190 thousand years ago were woolly mammoths.

The story of the mammoths should be a familiar one to us by now. Evolution was taking place in stressed populations experiencing change. The early glaciations had their first, and most permanent, impact in the north-east, far away from the mild oceanic conditions of Europe. It is here that evolution was active, selecting innovations that enabled survival as core habitats vanished. Because conditions worsened later on these new mammoths found that their habitat expanded, and they thrived. By the time they were reaching Europe, other mammoths in north-east Siberia had adapted to even harsher conditions and a new wave got under way. Each time it was the earlier versions that lost out, not necessarily through competition from new forms but because their world simply disappeared. In the north-east conditions were permanently and progressively harsh so gradual evolution was possible. In the south-west changes were too sudden and transitory for the local mammoths to adapt—here extinction and immigration were the rule.

The last woolly mammoths lived as recently as 4 thousand years ago in the steppe-tundra of Wrangel Island in the Russian Arctic. These mammoths were a dwarf version of their mainland counterparts,[18] possibly reflecting adaptation to low or poor feeding habitat. On the mainland, the last mammoths had died out over $5\frac{1}{2}$ thousand years earlier in the Taymyr Peninsula.[19] It marked the end of a long process of range contraction: each time climate warmed up, mammoth habitat shrank and mammoths lost out, and each time colder conditions returned mammoths staged a partial recovery. In the end it became a question of numbers and, with no further cooling of note, they died out. The story of the success of the Eurasian mammoths is another example of being in the right place at the right time; conversely their long investment in adaptations that suited them in cold environments was the cause of their demise when change, this time in the wrong direction for them, was too sudden to allow evolution any room for manoeuvre.

The woolly mammoth is often portrayed as a member of a cold-adapted fauna that lived in Eurasia during the Ice Ages, the glacials that we have been looking at in this chapter. Other animals that typified this fauna included woolly rhinoceros, reindeer, musk ox, and arctic fox.[20] The warm interglacial fauna has also achieved an identity of its own,[21] much as the cold fauna has, and both have become fixtures in the scientific literature. We saw in Chapter 3, however, that the term fauna has limited application as individual species have each tended to respond to climatic and environmental conditions in their own particular way. Emphasis on grouping animals into particular faunas has oversimplified the ever-changing coming and going of species during the glaciations.

The changing geographical distributions of animals during the glaciations were responses to a variety of circumstances. Rarely was climate directly responsible although the heavy woolly coats of mammoths, rhinos, and musk oxen undoubtedly had a heat conservation function. More often than not what drove the changes in areas lived in were shifting supplies of food, which were themselves controlled by climate. Because each species had different requirements and tolerances no pattern replicated itself exactly in more than one species. To further complicate matters, much depended on the location and state of the populations of each species when change came. We have just seen how mammoths in north-east Siberia and in Europe responded differently to changing circumstances. Mammoths also survived a number of periods of global warming in Siberian refuges but they eventually succumbed, showing that the outcome of change depended very much on the particular circumstances of each population at the time of change.

Although the distinction between cold and warm faunas has served to generalize about the way in which animals responded to the ebb and flow of the successive Ice Ages that hit Eurasia, a great amount of useful information has been lost along the way. The impression that I get when I read many descriptions of the alternating cold and warm cycles is of a Eurasian landscape that changed from closed forest, during warm and wet times, to open steppe-tundra during cold and arid times. Warm and cold faunas followed. But if we take a closer look at the published climate curves we see that the warm/wet and cold/dry periods actually took up a small amount of the total time of any glacial–interglacial cycle.[22] The greater part of the time was taken up by climate somewhere between these two extremes. Climate variability became more pronounced as we got nearer to the present day and the ecological turmoil became greater than before. It is difficult to visualize how, under these constantly changing conditions, closed forest or tundra-steppe could have ever blanketed the whole of Eurasia for very long. Even when full interglacial conditions took over, local geology, the action of large herbivores, and natural fires maintained a mosaic of closed forest, glades, savannah-type woodland, shrublands, and grasslands across much of north-western Europe.[23]

If we take a closer look at the fauna of Eurasia in the Middle Pleistocene, we are hard-pressed to find animals exclusive of dense forest. In fact only an extinct species of tapir, a forest herbivore with surviving cousins in the rain forests of south-east Asia and South America, just made it to the very beginning of Middle Pleistocene of Eurasia.[24] The climatic cycles that started 780 thousand years ago effectively removed the last of the warm and moist forests of Eurasia. Instead, what we find after this are various kinds of broadleaved and coniferous woodlands that never fully dominated the landscape. The mammals that thrived from the Middle Pleistocene onwards were mostly species at home in a range of habitats, a number of which included trees, but flexible enough to cope with a variety of situations from dense woodland with grassy glades to shrublands, forest edges, savannah-type habitats, and the steppe-tundra. Their success depended on the availability of plants that they could eat and digest efficiently. There were browsers of leaves and forbs and there were grazers—many of them were able to do both.[25]

Some animals required very specific conditions. They were a minority by the Middle Pleistocene and went extinct. The hippopotamus, a grazer, spread northwards into Europe as far as the British Isles but not very far eastwards on the continent where winter temperatures were a serious impediment.[26] This is an animal that does not like long, hard frosts and is at home in areas with mild temperatures and high rainfall. It is tied to living close to lakes and rivers so it would never have been widespread. Hippos were gone from the European scene by the last warm interglacial, around 125 thousand years ago, or shortly after.[27] The European water buffalo was another species that needed similar conditions to those of the hippo and also disappeared at about the same time.[28]

The Barbary macaque was one of the few monkeys to spread successfully away from the tropics.[29] Barbary macaques reached north into the British Isles and Germany and lived in forests across much of central Europe during mild interglacials. The Barbary macaque's origins in Europe date back to the warm Pliocene around 5 million years ago, well before the start of the glaciations, and the last ones disappeared, like the hippos and the water buffaloes, during the last interglacial.[30] The Middle Pleistocene populations of European Barbary macaque are best seen as relics of a warmer past that clung on in European refuges during repeated glacials but finally succumbed with the start of the severe conditions of the last glacial cycle.

The hippo, the water buffalo, and the macaque are three extreme versions of plant-eaters that were killed off by the successive waves of extreme cold that affected Europe after 780 thousand years ago and their favourite foods ran out. There were many other herbivores that also suffered when the cold, treeless, steppe-tundra overwhelmed most of the continent. Their stories are broadly similar although they differ in the detail. These animals ended up in local refuges where they rode out the bad conditions. They expanded from these refuges when conditions improved and woodland and savannahs regained a foothold. After repeated population crashes some simply could not recover and went extinct. Each species tells its own story so the pattern of extinction is not identical—not all species went extinct at the same time. Few, like the hippos, water buffaloes, and macaques, made it past the last interglacial. Some that did, like the narrow-nosed and Merck's rhinoceroses, and the straight-tusked elephant, did not make it much further after that either.[31]

The distribution of these rhinos and the elephant makes interesting reading and reveals a pattern that has striking parallels with that of their contemporaries, the Neanderthals. When conditions were mild and their preferred broadleaf woodland habitats expanded these herbivores stretched widely, from the Iberian Peninsula and the British Isles in the west, right across the wooded plains of Europe and Siberia almost to the shores of the Pacific in the Russian Far East. When woodlands shrank the range of these large herbivores became fragmented. The last populations seem to have made it past the last interglacial and into the start of the last glacial but by then the survivors were restricted to the Mediterranean, especially the mild Iberian Peninsula, where the last ones held out for a while before becoming extinct.[32]

The Neanderthals, and their *H. heidelbergensis* ancestors, were not the cold-adapted humans of Ice Age Europe as they have often been portrayed; instead they grew up in the woodlands and savannahs, often around wetlands, of the mid-latitude belt of Eurasia alongside these large herbivores.[33] These were highly rich and productive environments that offered a wide range of foods and opportunities for collecting them. At times when mild conditions allowed, they tracked these habitats northwards and reached the British Isles and Germany. Like the large herbivores of the temperate woodlands, further east their range was restricted to the southern fringe between the mountains of eastern Europe and Asia to the south and the treeless Russian plains to the north.

Homo heidelbergensis would have been at home in varied landscapes of trees and water but not in dense, impenetrable, jungle. In these environments these people would have been familiar with the animals of the temperate woodland mosaics, many of which would have been very large. Unlike the animals of the open plains, these ones probably did not live in large herds. They would have been, instead, widely dispersed across the landscape but they would have accumulated at higher density around favoured places such as sources of fresh water where grassland and woodland animals would have come together. It comes as no surprise that many of the *H. heidelbergensis* sites across north-western Europe, in places like Boxgrove, Hoxne, and Pakefield in the British Isles or Mauer, Schöningen, Miesenheim, and Bilzingsleben in Germany, are lakeside or riverside sites close to grassland and open woodland.[34] These sites were places of active hunting, scavenging, and butchery of large mammals. In Schöningen, beautifully preserved 400-thousand-year-old wooden spears testify to the hunting prowess and techniques of these people and show that wood was a raw material used by them.[35]

We saw in Chapter 1 that the people of the Pit of the Bones, a population of *H. heidelbergensis*, were well built, large-brained, and probably able to communicate with speech. The publicity that we have given to the Neanderthals over the years has undermined *H. heidelbergensis*'s achievement. Yet here we have a people at the height of physical accomplishment, powerful and intelligent hunters capable of despatching big mammals, and able to carve their own niche in a world full of large and dangerous predators. They were at the peak of their game in a rich and diverse world of mega-mammals. As this world began to disintegrate following wave after wave of cold and dry climate, species started to disappear as we have seen. The evolved *H. heidelbergensis*, whom we call Neanderthal, had entered a dying world, one of decay and ruin. As things got progressively worse this world lost more and more species, animals that were not replaced by new kinds.

By the time the classic Neanderthals had emerged,[36] during the last interglacial around 125 thousand years ago, they were already a people doomed to extinction. Like the hippos, rhinos, and elephants of the Eurasian forest, the Neanderthals were a population of living dead, existing on borrowed time. Like these other mammals, the Neanderthals had a short-lived moment of reprieve as the climate warmed up. The next time the climate would be this generous, 100 thousand years later, Neanderthals, straight-tusked elephants, and narrow-nosed rhinos would no longer be. Only their fossils would remain, buried and waiting to be discovered by another human who would ponder the

nature of his own existence and presumptuously imagine his involve-
ment in the debacle of his distant cousin, the Neanderthal.

Not all the herbivores of the temperate woodlands, savannahs, and
prairies vanished. Some, as we have seen, made it into the cooling
part of the last glacial cycle but eventually, like the Neanderthals, they
succumbed. Others were able to manage a little longer and some, the
general-purpose species able to ride the bad moments, are still with us
today. They include the red deer, the wild boar, and the ibex. These
medium-sized mammals became the regular prey of the Neanderthals
across much of their geographical range as they hung on for survival
following the collapse of global warming.[37]

So after a long story of successful survival, a population of H. erectus
that had entered Eurasia over a million years ago found itself in the right
place at the right time. The changing climate generated, for a while,
conditions conducive to the existence of communities of large grazing mammals in these temperate environments. For a while this was
probably a better place to live than even the tropical savannahs to the
south where water, as we saw in Chapter 3, would have severely limited
humans. The outcome was the Eurasian version of H. heidelbergensis.
But these conditions were not to last indefinitely and the response to
climatic deterioration, isolation, and impoverishment of resources was
the Neanderthal. For a while Neanderthals did quite well but conditions
did change and they found themselves in the wrong place at the wrong
time. Nobody could have predicted the changes in this volatile world
where extinction had overpowered evolution.[38] The Neanderthal was a
victim of circumstances.

Each time that conditions became colder and drier, and this was
to become the norm after the last interglacial, treeless environments
encroached into the wooded habitats that were the mainstay of
the *heidelbergensis–neanderthalensis* economy. That economy had been
based on the hunting and scavenging of large herbivorous mammals,
at least across much of the geographical range. We will see in Chapter 7 how the Mediterranean Neanderthals differed from their northern cousins by subsisting on a much more diverse range of resources.

H. heidelbergensis used spears to hunt mammals as we have seen and the Neanderthals followed the tradition. Ambush hunting, using thrusting spears, seems to have been a technique regularly used by them.[39]

There were two requirements to make ambush hunting successful. One was cover, which would allow the Neanderthals to get close to their prey. This was amply provided by the open woodland and savannah habitats in which they lived. The second was power, and they had plenty of that in their robust and muscular bodies, the legacy of *H. heidelbergensis*. *Homo neanderthalensis* was the product of several hundred thousand years of investment in the ambush hunting tradition within the temperate woodlands of Eurasia. The appearance of projectile technology among the Ancestors is often hailed as an advancement of technology,[40] but the reality is that such technology would have been useless against the powerful animals of Middle Pleistocene Eurasia. To down such animals needed power, guile, cooperation, and getting up close. That they did not shy away from contact is demonstrated by the injuries which they regularly sustained, comparable to those of modern Rodeo athletes.[41] Neanderthals would have often stared at their prey right in the eye.

So many thousands of years of overinvestment in a body capable of handling large mammals came with a price tag. The penalty was the inability to survive where there was no cover or where long-range movements to find herds were needed. In these situations body mass was an impediment. When cold pushed the tundra south and aridity pushed the steppe west, a new environment was created—the steppe–tundra. A new set of animals appeared on the scene. These animals included the woolly mammoth, woolly rhinoceros, musk ox, reindeer, and the Saiga antelope and they thrived in the expanding treeless environments that swept across Eurasia right down to France and northern Iberia.[42]

A Neanderthal would not have found these animals particularly daunting, especially the smaller ones. After all, at the level of food resource there was probably not much difference between a fallow deer, a red deer, and a Reindeer. The crucial difference was access. Whereas fallow and red deer could be stalked and ambushed, the Neanderthals would have stood out at a distance on the steppe–tundra. Getting close to a herd of reindeer was a different kettle of fish altogether. It comes as no surprise that the Neanderthals, or their predecessors, never ventured into the tundra or the steppe during the warm interglacials. The frontier of Neanderthal territory was defined where trees became so thin on the ground that ambush hunting became uneconomical or downright impossible. The problem facing the Neanderthals was that as climate became colder and drier, the frontier got perilously close to core territory. The change from wooded to treeless, when it came, was rapid and the Neanderthals had no option but retreat.[43]

This pattern of advance and retreat repeatedly isolated and reunited Neanderthal populations.[44] We have no knowledge of the size of the Neanderthal population at any stage but their presence in northern sites was limited to the warmer periods. The successive cold pulses gradually wore down their numbers and the repopulation during warm intervals does not appear to have allowed a full recovery. Each cold pulse affected a smaller and more fragmented population than the previous one until one day so few were left that recovery was impossible. Extinction followed.

As with the straight-tusked elephant and the narrow-nosed rhinoceros, the fragmented populations survived in the Iberian Peninsula, the Balkans, Crimea, and the Caucasus; these were areas of relatively mild climate and rugged landscape where pockets of woodland survived.[45] One such place was the northern shore of the Strait of Gibraltar, in full view of the coast of North Africa. This was a refuge for warm-loving plants that had become extinct everywhere else in Europe. Many species of reptiles, amphibians, and other animals intolerant of frosts, cold temperatures, and drought held out here. It was here, in a cave now known as Gorham's Cave on the Rock of Gibraltar that the last population of Neanderthals survived.[46] We will examine the different opinions of the causes of the Neanderthal extinction in Chapter 7.

If we were to take a snapshot view of the human world at 45 thousand years ago, on the verge of the entry of the Ancestors into

Europe, we would find scattered populations of Neanderthals living in pockets of open woodland and rugged landscapes along the southern fringes of the continent and eastwards along southern Siberia. These were populations under stress and in decline. The Neanderthals had been long gone from the Middle East. Africa was populated by the Ancestors who had by then spread right across the continent, eastwards across Arabia and India, and into China, south-east Asia, and Australia. A few isolated groups of *H. erectus* and *H. floresiensis* held out in remote tropical outposts of south-east Asia.

Today, animals like the red deer and the wild boar are remnants and reminders of the rich world that was once the domain of the Neanderthals and their ancestors. Irrespective of the position that we might take regarding the causes of the extinction of the Neanderthals, it is undeniable that by the time the Ancestors reached their strongholds in southern Europe and Asia these ancient peoples of Eurasia were already on the way out. The re-peopling of Eurasia was to be the inheritance of a new population, one that luck would bring to those in the right place at the right time.

6

If Only . . .

W<small>HEN</small> I first got involved with our origins back in 1989, the gospel went something like this: anatomically modern humans,[1] our ancestors, descended from a common ancestor— 'mitochondrial Eve'—who had lived 200 thousand years ago in tropical Africa. They emerged from this home and conquered the world. It was beyond question that these moderns had replaced each and every other population of human on the planet as they swept across the land. To even consider the merits of the alternative, multiregional, model made you an instant heretic.[2] Different versions tried to explain how the replacement happened. In its most extreme edition the story was fairly recently described as 'the modern human race's first and most successful deliberate campaign of genocide'.[3] This may make great headline material but where is the evidence of such genocide? There is none, just as there is no evidence of any kind of Ancestor competitive superiority over any other people that they came across. We saw in Chapter 3 how Neanderthals and proto-Ancestors were both present in the Middle East at a similar time even though we could not be certain that they had met. What we do know is that the proto-Ancestors disappeared and the Neanderthals remained and that, in itself, should raise some alarm bells regarding their predicted superiority. Could this presumption of superiority be wrong?

The idea of a single origin of the Ancestors is not new. It dates to the 1950s and was revived in the 1970s under the term 'Noah's Ark' hypothesis.[4] At that stage the region of origin was unknown but that became clearer a decade later when Rebecca Cann and colleagues published a ground-breaking paper which analysed the mitochondrial DNA of 147 living people from five geographic populations.[5] The results pointed to an African homeland 200 thousand years ago and the riddle appeared resolved. The paper was published on the first of January in 1987. Just over two months later a conference organized in Cambridge University brought together fifty-five leading researchers in the field of human origins. The first paper in the volume that was later published was by Stoneking and Cann. The title of the conference volume—'The Human Revolution'—gave a clear signal to all and sundry of how our evolution was to be seen from then on.[6] The subsequent two decades have seen a profusion of papers and books dedicated to supporting the human revolution. Some have been important contributions but many have been uncritical, unsupported, often subjective, accounts whose aim has been to explain the available evidence in favour of the Noah's Ark hypothesis.[7]

Many would regard the multiregional alternative hypothesis as defunct.[8] Multiregionalism's essence is the colonization of the Old World by *Homo erectus* and subsequent evolution of the different populations into the modern races of humans. It rejects the later expansion and full replacement of all human kinds by a single population of humans. Its origins go back to the 1920s when a Neanderthal phase of man was considered intermediate between *H. erectus* and the Ancestors.[9] Howells coined the term 'Candelabra' theory, the different branches representing evolution of different humans, a name synonymous with the multiregional theory.

Ironically, not Weindereich, Coon, or Howells, the major exponents of this view, are the names that appear when the debate rises from the ashes from time to time. Others have taken over the cause with renewed passion and ardour. Multiregionalism has had some very strong advocates, among the most notable being Loring Brace,[10] Alan Thorne, and, especially, Milford Wolpoff.[11] Chris Stringer is most often recognized as the standard-bearer for Out-of-Africa 2, being the presumed second major expansion of *Homo* from Africa with the first having been that of *H. erectus*.[12] And so the fiery Out-of-Africa 2 versus multiregional evolution debate became a crusade between the knights of Stringer and Wolpoff, one that had an added social importance as multiregionalism proposed a deeper divide in time between living human races than the Out-of-Africa 2 did. The embers still fly from time to time in palaeoanthropological meetings where our origins come under renewed scrutiny.

I have given this brief introduction to the contrasting theories here, before going on to give my own views, so that these can be seen against the historical backdrop. Despite the many advances, particularly in genetics, and new discoveries of fossils since the early days of the debate I remain unconvinced about the absolutism of the Out-of-Africa 2 replacement theory. This should not be taken to mean that I favour the extreme form of multiregionalism—I do not; but I do consider that the picture of human interactions across Eurasia, and elsewhere in the Old World, in the period between 50 and 30 thousand years ago was far more complicated than that of a simple replacement of one group of people by another.[13] It seems that I am not alone in thinking this way. Describing an early modern human fossil from Tianyuan Cave in Zhoukoudian (China), palaeoanthropologist Erik Trinkaus and colleagues recently commented that the fossil's anatomy 'implies that a simple spread of modern humans from Africa is unlikely'.[14] Yet, despite the many advances of the past decade that have cast a long shadow of doubt on the idea of a burst of human creativity that led to the human revolution,[15] the volume of papers of the follow-up conference to the 1987 Cambridge meeting, held in 2005, went by the title 'Rethinking the Human Revolution'.[16] Old habits, it seems, die hard.

I propose that we start with a clean sheet. We will take a look at the period between 50 and 30 thousand years ago: at 50 thousand Eurasia was occupied only by Neanderthals; by 30 thousand they were all but gone and the land mass was inhabited by the Ancestors. Let us for now forget about Neanderthals and Ancestors and, instead, have a look at the panorama across Eurasia during this time. What was the climate like? What environments were found where? What animals lived where? Where were the people? Once we have established the answers to these questions we can return to issues related to how humans got to where they did, who they were, and where they came from. Let us start with climate.

The climate of Eurasia partially recovered after a period of global cooling between 74 and 59 thousand years ago. At 50 thousand years ago climatic conditions were mild and relatively stable though not as warm as during the previous interglacial. The downhill climatic trend resumed around 44 thousand years ago and culminated in a peak of low temperatures 37 thousand years ago. The cold phase (27–16 thousand years ago) which culminated in the last Ice Age followed shortly after.[17] These trends were not uniform and were punctuated by warm events and very cold events at smaller time intervals, usually of the order of hundreds or thousands of years.[18] The frequent changes from warm to extreme cold and back made the period between 50 and 30 thousand years ago one of high climatic upheaval. The backdrop was steady climatic deterioration building up to the height of the last Ice Age that would follow soon after.

The climate changes were broadly similar across the entire geographical area of Eurasia,[19] but their impact on the environment varied significantly between regions of this huge land mass. And there are some surprises and controversies along the way: for example, some scientists interpret the environmental evidence as indicating that the extreme north-east of Siberia, where it faces north-western Alaska, was relatively ice free at the height of the last Ice Age.[20] Here the ice sheets were much smaller than they had been in earlier cold periods. One explanation for this unexpected observation may be that the Scandinavian and Barents/Kara ice sheets further west were so strong that they reduced the input of moisture reaching further east. Other scientists, however, support the idea of a much more continuous and extensive circumpolar ice sheet which reached a maximum area of 39 million square kilometres.[21] One dramatic consequence of such a huge ice mass covering much of Arctic Russia would have been that the exit of the major rivers that today flow into the Arctic Ocean from Siberia were blocked by huge ice dams. As a result massive freshwater inland seas developed across Siberia. When the ice dams became unstable as climate changed, vast areas suffered cataclysmic superfloods.[22] Such events would have caused untold damage, changed environments overnight, and decimated populations of many animals, including humans. We may never fully appreciate the impact of such glacial catastrophes.

These massive inland lakes would have been important barriers across a huge corridor of flat land that stretched from the Pacific coast of eastern Siberia, west all the way to the British Isles. This corridor would have been bound to the north by the ice sheets and to the south by the mountains of the Mediterranean, south-western and central Asia.[23] The east–west corridor changed in character in response to the whims of climate. The overarching trend of climate deterioration between 50 and 30 thousand years ago favoured the expansion of treeless habitats far and wide across this belt of plains. With increasing cold, tundra moved south ahead of the advancing ice sheets; with increasing aridity, steppe spread from its strongholds in the centre of Eurasia. Where they met a new habitat, the steppe–tundra was born (Figure 9). It was a double expansion wave that swept east and west across this vast area. When it got warmer and wetter, the two habitats decoupled and gave way to trees. Moments of reprieve for wooded habitats got rarer and rarer, especially deep within the continent where warmth and moisture became scarce environmental commodities.

We saw in the previous chapter that the mammals of the woodlands and savannahs had thrived across this belt for a long time.

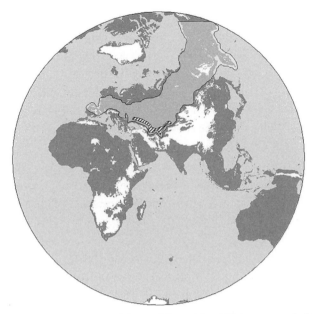

Figure 9 The steppe–tundra of the Middle and Late Pleistocene, a habitat first colonized by a population of *Homo sapiens* from the hatched region

Homo heidelbergensis and Neanderthal had been at home here too. Now climate was dealing blow after blow to these animals by reducing their favourite grazing and browsing woodlands to isolated remnants. Other mammals, like the woolly mammoth, had been slowly evolving in areas with permanent treeless landscapes and it was now their turn to colonize large areas of Eurasia. This world was ready for animals which could thrive by grazing in treeless habitats. Those that could not detach themselves from a browsing existence on leaves were on their way out along with the predators that relied on ambushing them from cover.[24]

So who benefited from the environmental tragedy that befell Neanderthal, straight-tusked elephant, and leopard? One group was made up of the herbivores of the cold north: the woolly mammoth, woolly rhino, reindeer, and musk ox. Another was of those of dry steppe and prairie: horse, steppe bison, and saiga antelope. For these species the period between 50 and 30 thousand years ago marked their apogée, reaching far west into France, the British Isles, and northern Spain. Then there were species sufficiently flexible in the food that they ate to get by: moose, giant deer, roe deer, red deer, aurochs, chamois, wild sheep, and ibex.

The large carnivores that survived were those also able to eat plants and fruit and hibernate during the cold, dark, winters: brown and cave bears. The other carnivores were smaller species that did not depend on the large herbivores. The arctic fox reached right down to France and the red fox retreated. Lynxes and wild cats completed the panorama. But it was the predators that could follow the large herbivores across the steppe-tundra that thrived. Lions and spotted hyaenas managed to make a living for a while but they were no match for the ultimate long-distance pursuit runner.

The wolf became the master of this open landscape. Only in the far north did it have a rival: the polar bear had evolved from a population of brown bears and became a specialized meat eater.[25] Like the wolf, it ranged across great distances in pursuit of its prey but, unlike the wolf, it went to sleep at the height of the northern winter. The wolf and the bears together give us a picture of what was needed to be a successful hunter in these landscapes. Most important of all you had to be a marathon runner; energy-sapping sprinting was not a viable alternative. The ability to mix the diet, to ride bad periods by storing reserves as fat and reducing energy expenditure or by caching food, and hunting cooperatively in groups were optional extras. A hunter that could do a number of these things had the potential to become the super-predator of the treeless plains of Eurasia. We must wait until Chapter 8 to find such a character.

It is time to see where the humans were during this period of climatic and environmental turmoil. Ironically, we know much more about the Neanderthals than we do of the Ancestors. The Neanderthals 50 thousand years ago had already experienced the effects of the encroaching treeless habitats across the Russian Plain and eastern Europe and their range had started to contract. By 40 thousand years ago their homeland had been pinned back to the Mediterranean, south-west

France, and pockets round the Black Sea.[26] The acceleration of cold and unstable conditions after 37 thousand years ago reduced the range even further,[27] leaving a major stronghold in southern and eastern Iberia and pockets in northern Iberia, the Atlantic seaboard to the north, in the Balkans, Crimea, and the Caucasus. By the time we reach the end of our period, 30 thousand years ago, the only Neanderthals left were in south-western Iberia.[28]

The elusive Ancestors are harder to find. We get glimpses of these people in eastern Europe, to the north and west of the Black Sea, between 36 and 30 thousand years ago.[29] These are the earliest Ancestor remains anywhere across Eurasia and the dates are only surpassed by specimens from Nazlet Khater in Egypt, dated to around 37.5 thousand years ago,[30] Niah between 41 and 34 thousand years ago, and Lake Mungo around 42 and 38 thousand years ago (Chapter 5). The available genetic evidence complements this picture and suggests that prior to 30 thousand years ago pioneer groups of Ancestors (and perhaps also proto-Ancestors?) barely penetrated northern Eurasia and were unable to gain secure footholds once there.[31] Our knowledge of their ecology is virtually nil.

The archaeology paints a very different picture. In Chapter 3 we saw how Neanderthals and proto-Ancestors in the Middle East made similar stone tools around 130 to 100 thousand years ago. In the absence of human fossils there was little that we could do when faced only with stone tools. The technology employed belonged to the Middle Palaeolithic family and went by the name of Mousterian. It seems that the Neanderthals kept on doing the Mousterian across Eurasia well after this. All Eurasian Mousterian sites with fossils between 120 and 28–24 thousand years ago are exclusively associated with the Neanderthals so it is safe to assume that they give away the presence of these people. The problem has been with the array of novel technologies, under the generic banner of Upper Palaeolithic, which started to appear in Eurasia around 45 thousand years ago. The traditional interpretation has been that these were the work of the Ancestors and so our entire understanding of their arrival in Europe has been based on the use of technology as proxy for the biological entity that was our direct ancestor.[32] But can we really be so sure?

The surprising thing is that the only certainty that we have about the makers of the many stone tool cultures that appeared in Eurasia between 50 and 30 thousand years ago is that the Mousterian was made by the Neanderthals. We also have a good idea that the transitional culture that appeared in France around 45 thousand years ago and persisted until 36.5 thousand years ago,[33] known as the Châtelperronian, was also made by the Neanderthals,[34] but the association between tools and fossils is so limited that we should not discard the possibility that other humans were also linked with this culture. Other transitional cultures appear across central and eastern Europe and the Middle East during this period but, so far, none have been linked to human fossils.[35]

Until recently the earliest Upper Palaeolithic culture in Europe, the Aurignacian, appeared firmly linked with the Ancestors. It was the distribution of this culture across Europe that appeared to confirm beyond doubt the spread of the Ancestors from the Middle East.[36] The Aurignacian was associated with Ancestor remains and the classic site for this was Vogelherd in Germany. But, then, in 2004 the results of direct radiocarbon dating of the human skeletons at Vogelherd threw a spanner in the works. As it turned out the skeletons were not associated with the Aurignacian artefacts at all; the human remains had been buried into the site much later, during the Neolithic between 3.9 and 5 thousand years ago.[37] When we take stock of the Aurignacian, and the claims for association with Ancestor remains, we can only conclude that we do not know who was responsible for this culture either. We cannot exclude the possibility either that, as with the Châtelperronian, Neanderthals and Ancestors shared this culture. We simply do not know.

So what can we take from this frustrating picture of gloom and doom? We should not despair. In recognizing the deficiencies and gaps in our knowledge we can, at least, try and see what it is that we can say about the picture across Eurasia between 50 and 30 thousand years ago. We have clear evidence of a progressive Neanderthal retreat as the steppe–tundra advanced. We have genetic evidence of feeble pioneering forays of groups of Ancestors into Europe, probably from the Middle East. We also have fossils that confirm their presence in central and eastern Europe from 36 thousand years ago. We observe the shrinking of the Mousterian culture as the Neanderthals retreated and we have the flourishing of many new cultures that come under the umbrella of transitional or early Upper Palaeolithic. All this points to a period of climatic and environmental chaos, during which time human groups, of unknown biological identity, were trying to deal with the upheaval by trying out new things.

There was no clear-cut outcome in terms of superiority of one kind of human over another or of one culture above the rest. The great cultural diversity across Eurasia also tells us that there must have been prolonged periods of isolation between regions that maintained the separate identities of the peoples of mid-latitude Eurasia. Only the Mousterian and the Aurignacian appear to retain a wider geographical spread,[38] though limited to particular ecological conditions. This should give us a clue as to what was going on.

The presence of transitional and early Upper Palaeolithic cultures in the Middle East around 45 thousand years ago has for a long time been taken as support for the Out-of-Africa 2 model. Here we could see the cultural change into something new and supposedly modern. It was the start of the wave of advance of the Ancestors from Africa. We have seen how it is not possible to take these cultures as confirmation of the presence of the Ancestors. In any case there is clear evidence now that such cultures were appearing across Eurasia at about the same time. We have seen this for the Châtelperronian in France and the same is true for the transitional and early Upper Palaeolithic cultures of central and eastern Europe, the plains north of the Black Sea, and east across southern Siberia to the Altaï Mountains.[39] So, rather than the spread of the Ancestors what the cultural evidence shows is widespread experimentation and innovation across Eurasia and the Middle East at about the same time. It is no coincidence that this was the time of climatic downturn when conditions were reaching the height of unpredictability.

We have seen throughout this book how biological innovation has been most active among peripheral populations, those living on the margins of others living in core areas. The people living in Eurasia and developing transitional or early Upper Palaeolithic cultures were, not surprisingly, those on the edge of the geographic range.[40] As the steppe–tundra encroached, these edge populations became the front line troops. They were faced with two options: adapt to the new circumstances quickly or die. The required change involved finding ways of living in, and hunting the animals of, the alien open and treeless habitats suddenly appearing everywhere.

An example will illustrate how people were attempting to cope. The Vézère Valley in south-west France was a contact zone between woodland habitats and steppe–tundra between 34 and 27 thousand years ago.[41] The people who lived in the valley belonged to the Aurignacian culture. Biologically, they may have been Ancestors but we cannot fully discard the possibility that they might have been proto-Ancestors or even Neanderthals. The early part of the period was typified by a cold and dry climate. A cold steppe dominated in which the Aurignacians hunted primarily reindeer followed by horse. They roamed widely across the open lands to track the herds and they made tools out of blades that were easily transported from one place to another, using raw materials often procured from long distances. The latter part of the period was warmer and wetter and wooded savannah and forest took over. The landscape had become a mosaic of habitats with a greater diversity of mammals so the Aurignacians did not to have to travel so far to find food. At the same time reindeer had become scarcer than earlier as their habitat retreated. So the Aurignacians modified their behaviour, becoming more sedentary and exploited red deer, wild boar, and aurochs in particular. Raw materials to make stone tools were procured locally, suiting the more sedentary lifestyle, and the types of tools made also differed from the earlier period. It seems that the Aurignacians behaved with great flexibility and adjusted their activities and tools in response to changes in the environment.

The kinds of changes that the Aurignacian people were experiencing in France were typical of the unstable world of northern Eurasia during this time. An excellent and dramatic example of these changes is provided by the pollen record from the Lago Grande di Monticchio in southern Italy.[42] The pollen record spans the past 100 thousand years and so includes the period between 50 and 30 thousand years ago. The site records constant and rapid shifts in the landscape around the lake, from steppe through wooded steppe to forest and back. It is precisely these kinds of changes that were happening across many areas and to which the people of Eurasia had constantly to adapt and re-adapt. The most striking insight provided by Monticchio is the speed of change as the major vegetation changes were taking 142 years on average. This means that one generation of humans might have been living in a forested landscape, their children in wooded steppe, and their grandchildren in open steppe. Because the changes were not just in one direction, later generations might have lived once more in wooded steppe or forest.

These changes were most frequent and intense in zones of contact between the plains of northern Eurasia and the hills and mountains of the south. These zones of contact were often mosaics of different habitats over small areas, as in France. During warm intervals forests would grow in the lowlands and up the slopes to the tree line. As conditions became colder the tree line descended and the forests remained in isolated and sheltered valleys. If these conditions persisted, the woodland might disappear altogether. Conditions away from these contact zones would have been more stable. We see this across central and south-eastern Europe. At this time the climate of central Europe was cold but stable and the dominant mammals were those of the steppe–tundra, especially woolly mammoth and reindeer.[43] These animals were scarce or absent in the Balkans to the south where there was a much greater variability in the animals present, which included species of woodland such as aurochs, red deer, and wild boar.

This ecological variability continued along the edge between plains and mountains right across to south-eastern Siberia. As we travel further east and north we enter the heartland of the steppe–tundra and it would have been in the plains of Central Asia and Siberia that the treeless habitats and its characteristic animals would have been most permanently established. This was their stronghold. The southern fringe, with mountains reaching up to 3 thousand metres in height, would have been the zone in which the woodland and open landscape animals would have been in close proximity for the longest intervals of time, much more than in the west. The typical habitats here alternated between forest of different types, wooded steppe, forest tundra, and steppe–tundra.[44] Woolly mammoth, woolly rhino, horse, wild ass, camel, wild sheep, wild goat, giant deer, reindeer, moose, saiga antelope, and steppe bison were among the dominant herbivorous mammals and the carnivores included lion, lynx, wolf, and brown bear. This diversity is in complete contrast with the small range of mammals in the treeless expanses of the Arctic zone of northern Siberia,[45] well away from the plains-mountains contact zone.

The kind of brusque ecological changes that we saw in Monticchio were not unique. They typified the contact zone at this time of huge climatic and environmental turnover. The Siberian Arctic was, paradoxically, stable though cold but humans only managed incursions here after 36 thousand years ago.[46] But the animals of the far north did come south and west as their habitat expanded and they came into contact with humans on the fringe areas of southern Siberia and west all the way to France. This edge between woodland and treeless habitats waxed and waned many times along a narrow belt between plains and mountains. This was the tension zone, in places like the Altaï, the Carpathians, and the Pyrenees, where humans were being stretched to the limits of their ingenuity and it is right across here that we see cultures and technologies that we define as transitional or early Upper Palaeolithic. This was the novelty sector where over and over again, for fifteen millennia, survival battles were won and lost.

The tension zone had another face. This one was south of the mountains, from the Middle East in the west to northern India in the east.[47] This face was also one of sharp ecological contrasts but these were of a very different kind from those to the north. Here too there were zones of sharp changes between wooded and treeless landscapes but the latter took the form of dry steppe and desert as we saw in Chapter 3. The animals of the steppe–tundra never reached these latitudes and the changes in potential prey involved gazelles, fallow deer, red and roe deer, aurochs, wild sheep, and ibex.[48] The kinds of cultural transitions taking place in northern Eurasia were also happening here as people were repeatedly being faced, not with the progress of the steppe–tundra, but with a potentially worse problem—the advance of the desert. Here too we have a problem identifying who was making what. It has been largely assumed that the people of the Middle East at this time were the Ancestors and that the Neanderthals had long since abandoned these lands. This may be true but the fossil evidence in this part of the world during this time is hardly categorical: a skeleton at Nazlet Khater in Egypt that we have seen has been dated to 37.5 thousand years ago; a child from Ksar 'Akil in Lebanon dated to 35 thousand years ago; and fragments of skulls from Qafzeh dated to between 30 and 28 thousand years ago.[49] As in Europe, there is no direct evidence of Ancestors in the Middle East prior to 38 thousand years ago and it is only much later, with the Kebaran culture after 20 thousand years ago, that we have a clear and unequivocal signal in this region.[50]

We saw in the previous chapter that the Ancestors, based on genetic evidence as human fossils were virtually absent, were in India by 50 thousand years ago and probably considerably earlier. From here we picked up a rapid expansion into south-east Asia and Australia. Following the genetic signal, we find that others dispersed north and west, instead of penetrating deep into India, at about the same time. They reached the Middle East and the northern and southern shores of the eastern Mediterranean. The genetic markers of this dispersal are not strong among present-day Europeans, and suggest that this movement into Europe was hardly a momentous event,[51] a fact borne out by the lack of human fossils. Attempts have been made to match the genetic signal with the archaeology, specifically the Aurignacian culture; but we have already seen that we cannot identify the makers of this culture which, in any case, appears to have been a European invention and not an import from outside.[52]

What can we conclude about the period between 50 and 30 thousand years ago when we put together what we have seen so far in this chapter? Climate became progressively inhospitable. Repeated and sharp oscillations did not allow particular habitats to become well established. Changes were most abrupt where plains met mountains and different habitats were in close proximity. Most people were concentrated along the edge between the mountains and the plains, in places of high ecological diversity within small distances. These places offered a range of options for making a living.

Few people had managed to get further north, away from the contact zone and into the open plains. Humans had been living to the south of the contact zone for a long time and had responded to periods of aridity, when the desert encroached, by developing projectile technology that enabled them to switch from ambush hunting to long-distance tracking and hunting of gazelles and other desert animals.[53] This flexibility may have assisted their subsequent expansion. The human panorama across northern Eurasia between 50 and 30 thousand years ago was therefore one of colonization, extinction, and innovation. But was it also one of contact and conflict?

The decade that started in 1998 saw a heated debate between scholars in pursuit of an idea. As it turns out it was a sterile debate without hope of being resolved satisfactorily, for the simple reason that both sides failed to recognize the limitations of the evidence at their disposal. Archaeologists Francesco d'Errico, Joao Zilhão, and colleagues had latched onto a paper published in the journal *Nature* two years earlier which implicated the Neanderthals with the Châtelperronian culture.[54] The Châtelperronian was, as we have already seen, a transitional culture that incorporated Middle and Upper Palaeolithic elements and included a number of key features that had until then been considered to have been the exclusive handiwork of the Ancestors—worked bone, ornaments, etc. Now d'Errico, Zilhão, and colleagues argued that this was clear evidence that the Neanderthals were the sole makers of the

Châtelperronian and this was proof of their abilities, comparable in every way to those of the Ancestors.[55]

Anything else in the argument apart from the kind of extrapolation, from an association between Neanderthals and this material culture, to mean that only Neanderthals made it typifies many of the unsatisfactory conclusions reached from very little evidence. From accepting that Neanderthals had been able to make tools and ornaments comparable to those made by the Ancestors, an apparently sound conclusion, suddenly the jump was made to giving them the exclusivity over the Châtelperronian. It was a symptom of a long-standing, and in my view highly erroneous, observation among some archaeologists that equated biological entities, such as the Neanderthals, with particular cultural traditions. Funnily enough these same authors have argued later on precisely against this kind of straightjacket link between biology and culture.[56]

I organized an international conference in Gibraltar 1998 to commemorate the 150th anniversary of the discovery of the Neanderthal skull at Forbes' Quarry (Figure 10).[57] The Châtelperronian paper had been published and one of its authors, Joao Zilhão, was one of the guest speakers. Another was Paul Mellars, an archaeologist from Cambridge University. I had not anticipated the reaction to the Châtelperronian paper. It dominated the entire conference and it drew the battle lines between those who saw the Neanderthals as perfectly capable of behaviours that had previously only been attributed to the Ancestors and those who did not. Those who did not, Mellars at the helm, argued that the Neanderthals had either become acculturated by being in contact with the Ancestors or had traded the 'modern' objects with the newly arrived humans. Either way, they could not have made the artefacts without outside help. The battle between the two camps continues to this day.[58]

The entire discussion about whether Neanderthals met the Ancestors in western Europe some time between 40 and 35 thousand years ago has depended on the Châtelperronian culture being the exclusive domain of the Neanderthals and the Aurignacian of the Ancestors.

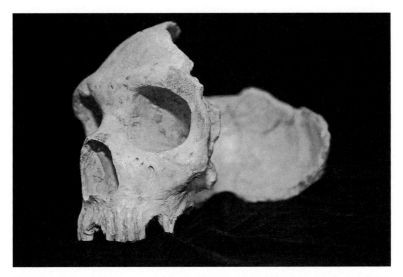

Figure 10 The Forbes' quarry skull, discovered in Gibraltar on the 3 March 1848, and pre-dating the Neander Valley discovery by eight years

We have already established that we do not know the identity of the makers of the Aurignacian and we only know that the Châtelperronian was made by Neanderthals from a few sites. For all we know the two cultures were made by one, other, or both, or indeed by proto-Ancestors. The observations of Aurignacian artefacts being found embedded amidst Châtelperronian archaeological layers, if true, need not mean that Neanderthals were interchanging or copying what they saw the Ancestors doing. It could even be that one group of humans switched the kinds of tools that they were making in response to a change in the environment, just as the Aurignacians were doing in the Vézère Valley. So the opposing constructs, that Neanderthals could make their own Upper Palaeolithic tools and ornaments or they got them instead from the Ancestors, are castles built on air.

These disproportionate labours aimed at teasing out a few tools and radiocarbon dates, often from museum objects that had been excavated a century earlier, from fine layers in a few caves have created a false impression that high science was being applied to the problem. Such intense efforts have placed so many trees in the way that the view of the woods has been totally obliterated. The reality may have been much simpler. Is it a coincidence that all the transitional and early Upper Palaeolithic cultures bordered the edge of the hills and plains? Is it a coincidence that away from these edges we do not find such cultures? The answer may lie in the ecology, but to be able to solve the mystery we will need first to know the function of the tools and other artefacts being made by people across northern Eurasia and the Middle East between 45 and 30 thousand years ago.

The behaviour of the Aurignacians in the Vézère during cold periods when the trees disappeared should give us a clue. They were making tools that were small and portable. These were often armatures placed on wooden shafts to make lightweight spears or arrows that could be thrown from a distance, probably with the help of spear throwers, at medium-sized or small animals. Because they were hunting animals over open landscapes they had to move around a lot more than in woodland. The price to pay with this strategy was increased energy expenditure in travelling in search of these animals but, once located, the prey was usually in herds and open to tactical group hunting. These bonanzas of protein were patchily distributed across the landscape and the key was how to find them. Once the herds were located, projectiles provided a means of killing from a distance, a useful tactic where cover was limited. Because this style of hunting required travelling across vast distances, often away from the sources of prime raw materials from which to make the armatures, these people selected high-quality materials and kept reusing the same artefacts. Recycling was invented by the early hunters of the plains of Eurasia.

A high-quality, portable, and reusable projectile weapon and tool kit was an essential feature of the first people who ventured away from the woods and savannahs of Eurasia.[59] It was a defining feature of all the various and diverse cultures that we have identified as transitional or early Upper Palaeolithic, be they in France, the Carpathians, or the Middle East. That some of these became widespread across large areas, for example the Aurignacian, suggests one of two things:

some technologies were more efficient than others and spread with their makers or they percolated from one group to another through interchange of ideas and trade. If it was the latter, we are left with the question of who acculturated whom? Why should it be that it was just the Neanderthals who picked new ideas from the Ancestors and not the other way round? Perhaps they each learnt from the other.

Unfortunately, this is a less clear-cut picture than one in which Neanderthals and Ancestors met and one imitated the other. But it is a picture that is likely to represent what actually went on all those thousands of years ago. People were suffering the pinch where it hurt—in their stomachs—and many would have died of starvation. South of the mountain belt, in the Middle East, Arabia, and across North Africa, drought would have been an added factor. Those who managed to find novel ways of coping with the stresses survived and may have recovered, when things improved during the many wild swings that the climate took. When this happened, and the forest returned, the portable projectile technology would have become obsolete. Just imagine having to throw spears at a deer when there were trees in the way. So they went back to the old tools and thrusting spears. Next time it got bad the traditions of making projectiles might have been lost and would have had to be reinvented. Elsewhere, the know-how might have survived and the weapons re-emerged. If this was all happening, and the climate record suggests that it was, at scales of human generations then what chance would we have of picking up the fine detail from the archaeological record?

The archaeological record points to people living in many areas of northern Eurasia between 50 and 30 thousand years ago but we have no real indication of their numbers. In all likelihood they would have been thin on the ground and extinction of local groups would have been the norm. It was, after all, what was happening to the Neanderthals and we have no reason to suppose that the pioneer Ancestors were any better at dealing with the situation. True, it seems that some penetrated into Europe and parts of northern Asia but Neanderthals also extended their area northwards during the milder episodes at this time. Which groups made it and which did not would have depended on a large dose of luck, on being in the right place at the right time (as we saw in the previous chapter) and on the way climate went. If only climate had become progressively warmer instead of colder, what story might others be telling today?

One niggling issue remains that we need to deal with before moving onto the next episode of the story. Did the Neanderthals and the Ancestors interbreed? Although this has been a long-standing question it received heightened importance in 1999 when a team led by Portuguese archaeologists made the bold claim that they had found a skeleton that showed precisely that Neanderthals had mated with the Ancestors.[60] The skeleton was claimed to be anatomically intermediate between a Neanderthal and an Ancestor and the discovery inevitably opened up a new controversy.[61] To date there is no consensus on the nature of this skeleton but what is, for me, unacceptable is the claim that the skeleton is an indication that Neanderthals and Ancestors mated regularly and in many places. This is yet another example of extreme overgeneralization but we should try to see the logic behind such an apparently rash claim.

The Portuguese hybrid child, assuming that is what it was, lived around 25 thousand years ago. At the time when the discovery was made the last Neanderthals were thought to have gone extinct around 30 thousand years ago, at least five thousand years earlier than the time when the hybrid child had lived. This was the key that led to the claim that Neanderthals had mated frequently and widely with the Ancestors. How else could we account for a hybrid so long after the last Neanderthals had died out? There must have been so much interbreeding around that we could still pick the traces up long after the Neanderthal extinction. The logic seemed sound although the evidence was clearly limited and circumstantial. Then, in 2006, I published a paper, together with a number of colleagues, in which we reported the late survival of Neanderthals on the Rock of Gibraltar, to between 28 and 24 thousand years ago.[62] This result placed the last Neanderthals in the same time frame and geographical area as the hybrid child and raised doubts about the claimed widespread hybridization of

Neanderthals with Ancestors. If it had happened the evidence for it would have to be sought elsewhere.

The other end of Europe, Romania, became the focus of attention for the hybrid crusade.[63] Discoveries of *Homo sapiens* remains, the earliest in Europe, were claimed to exhibit traits typical of the Neanderthals. These and other finds from central and eastern Europe collectively led to the conclusion that there had been modest levels of assimilation of Neanderthals into Ancestor populations as the latter spread into Europe.[64] Of course, all these results presuppose what a Neanderthal–Ancestor hybrid was expected to look like, having some degree of intermediate anatomy between one and the other. But hybrids are not necessarily expected to be like this. In baboons, hybrids of olive and yellow baboons are not intermediate between the parents. Instead the hybrid population is much more variable in features than either of the parent populations and often shows novel features not found in either of the parent populations.[65] So increased anatomical diversity and novelty at the level of the population, rather than individual intermediates, are the expected outcomes of hybridization. To find evidence of this is a practical impossibility when dealing with isolated and fragmented fossils and dates that have error margins that do not allow the kind of resolution needed to pin such a phenomenon down beyond any level of doubt.

The successful extraction of DNA from Neanderthal fossils opened another window into the past and the search for genetic evidence of interbreeding took a different route.[66] The results so far, while not excluding the possibility of genetic interchange altogether, strongly point away from it. Neanderthal genes seem to show that Neanderthals and Ancestors did not mix.[67] If we add to this clear evidence that anatomical differences between Neanderthals and Ancestors were deep-rooted, were already present in foetuses before birth, and were maintained throughout development,[68] the chances of viable hybrids making it to reproductive age would seem to be quite slim.

Put together, the anatomical, genetic, and developmental information at our disposal, coupled with the ecological picture of human populations thinly spread on the ground, would seem to point away from any kind of significant genetic mixing between Neanderthals and the Ancestors. If indeed they did mix, the evidence available to us today strongly favours the view that the Neanderthals did not contribute to our gene pool in any significant way. In any case we have seen that the humans that reached Europe at a time when there were still sufficient numbers of Neanderthals about left little trace of their genetic make-up among present-day Europeans. This leaves open another door which is that some Ancestors (and perhaps also proto-Ancestors?) and the Neanderthals did intermix but the trace was lost afterwards as Neanderthals disappeared altogether and the pioneer Ancestors almost completely too. To see what happened to these people we need to move forward in time to the period after 30 thousand years ago. But first we will pause and pay a visit to the last Neanderthals.

7

Africa in Europe—
A Mediterranean Serengeti

A line of cork oak trees was silhouetted against a pastel sky as the sun rose as a red ball over the low and distant horizon. Patches of orange water glistened against a dark background of sedge and rush. The nocturnal chorus of frogs sang their encore and went to sleep before the first hungry kites took to the wing in search of breakfast. Raucous calls from within the oaks signalled another day in 'la Pajarera', as the huge heronry was known in these parts. The tones became harsh as the lifting sun shone brightly across the marsh. Serenely sitting on the highest branch of the highest tree was the lord of this scene—the Spanish imperial eagle. This is how I recall the wilds of the National Park of Doñana in south-western Spain.[1] The Victorian naturalist Abel Chapman fittingly described this remote corner as a piece of Africa in Europe. But why am I describing this Garden of Eden at all in a book that is about our evolution and that of our Neanderthal cousins?

The scene that I describe could have easily been seen by a Neanderthal 30 thousand years ago. This is not a romantic extravagance of my imagination. It is statement based on solid evidence excavated over a period of seventeen years in the caves of the Rock of Gibraltar. It was here that the last Neanderthals on the planet survived for at least 2 thousand years later than anywhere else.[2] These caves hold a unique archive of information,[3] and have provided the clearest and most precise picture of the landscape within the territory of a Neanderthal group anywhere. The evidence includes pollen, charcoal from the Neanderthals' camp fires, the remains of the animals that they ate, and the tools that they used. They can be pieced together to form a unique and impressive jigsaw puzzle. Doñana is the closest match we have today of that landscape—if you like think of it as the image on the box of our brainteaser: it was coastal and sandy but this should not be taken to mean that it was sterile or inhospitable. On the contrary, the patchwork of habitats that the Neanderthals had outside their caves in Gibraltar was, like Doñana today, a rich ecosystem home to many different kinds of plants and animals.[4] I will start this chapter by painting a picture of what this Neanderthal landscape was like.

Gorham's Cave, the main but not the only source of the evidence, is like a huge multilayered cake (Figure 11). The massive cathedral-like cave stands out at the base of a tall cliff that rises 426 metres from the blue waters of the Mediterranean Sea. It is one of a line of caves that were all occupied by Neanderthals, a kind of 'Neanderthal City' in the southernmost part of Europe where it faces Africa just on the other shore of the narrow Strait of Gibraltar.[5] Only in the Middle East did the Neanderthals live further south. The cave is not a huge hollow though: it is filled with what at first sight looks like sand. Sand is certainly a big component of the cave's infill but it is mixed with bat guano, Neanderthal refuse, fallen stalactites, and anything else that may have found its way in over the millennia. In all, the infill is 18 metres thick. The base contains the oldest items and is dated to the last interglacial around 125 thousand years ago. You travel forward in time as you move up the cave, first to the last Neanderthals of 28–24 thousand years ago,

Figure 11 Rock of Gibraltar with Gorham's cave, where the last Neanderthals lived, at the base (second large cave from left). Above, present day with high sea levels; below, the predominant situation when Neanderthals lived in the cave, with sea levels lowered by up to 120 metres below present level, exposing a huge shelf where the Neanderthals foraged

Photo credit: Clive Finlayson; reconstruction: Stewart Finlayson.

then to the arrival of the first Ancestors around 20 thousand years ago, after this to the end of the Ice Age 10 thousand years ago, and above this, historical levels.[6]

Many of the fossil plants and animals identified from the various levels are excellent indicators of climate. Among the plants the olive, the stone pine, and the lentisc are evidence of a warm Mediterranean climate. Among the animals, the Hermann's tortoise needs a mean annual temperature of 14 °C for its eggs to hatch and it does not tolerate rainfall in excess of 700 millimetres per annum. Many species can be used in this way. Putting the climate tolerances of each of them for a particular level together allows us to say, with some confidence, what the climatic conditions outside the cave were when these animals lived there. The staggering conclusion when all the results were pooled together was that for the greater part of the past 125 thousand years the climate was almost the same as it is outside the cave today. At times it was slightly cooler and drier and at other times it was slightly wetter but overall it changed little.[7] These findings are hugely significant when we bear in mind the kinds of rapid and intense climate changes that, as we saw in the previous chapter, engulfed northern Eurasia, especially after 50 thousand years ago. It seems that the deep south-west of Europe remained largely unaffected by these changes. This conclusion was supported by the complete absence of any of the steppe–tundra mammals ranging widely further north. No woolly mammoth, woolly rhino, reindeer, musk ox, steppe bison, saiga antelope, cave bear, or arctic fox ever reached these latitudes. This was indeed another world, a piece of Africa in Europe as Chapman had proposed.

The plant and animal remains also provide us with very clear indications of the kinds of habitats outside the cave. For over ten years my wife Geraldine and I travelled the length and breadth of the Iberian Peninsula with one purpose in mind. We would sample the rich variety of habitats of Iberia, from the frozen wastes of the peaks of the Pyrenees to the warm Mediterranean woods of the south-west, from the humid forests of Cantabria in the north-west to the deserts of Almería in the south-east.[8] We would stop at particular locations and record the nature of the vegetation within an area of a hectare. Our measurements included the kinds of plants that we found and the structure of the habitat.[9]

We did this for one thousand plots, which gave us a fantastic database of information. Birds were easily observed during our work so we added estimates of the numbers of each species observed in each plot to our catalogue. We could work out, for example, the kind of habitat in which a particular bird lived? and also the limits of its tolerance today. Since most of these species had evolved before the Ice Ages and had survived them, their anatomy had been shaped a long time ago to suit particular habitats and ways of feeding. So, our logic went like this: if we found a particular bird as a fossil in the cave then it had to indicate that a particular habitat had existed outside the cave. If then we found a second species with similar habitat requirements our confidence grew. Our prediction would gain in confidence the more species with similar needs we found.

We then applied an independent estimate by looking at the plants themselves. Cork oak trees were different from Holm oak trees that were in turn different from Maritime pines. Each developed a particular habitat structure. So, if we had a range of birds and plants that indicated very similar habitats, we were confident that that particular kind of habitat had existed outside the cave at the time that the fossils had accumulated. If there was also evidence of Neanderthal activity in the same level in the cave it had to mean that the kind of habitat identified had existed when the Neanderthals had lived in the cave. Now we could start to say things about the environment of the Neanderthals. And by comparing levels that represented different time slices we could see whether the environment had changed at all and in what way.

To cut a very long story short what we found was that the Neanderthals who had lived in Neanderthal City had exploited a habitat mosaic outside the cave almost identical to what we knew in Doñana. There were shifting sand dunes, moving with the predominant winds and engulfing vegetation or enclosing it into 'corrales'.[10] There was a wooded savannah formed by stone pines, cork oaks, and junipers; there were thickets of denser vegetation, especially close to streams where willows and reeds formed impenetrable aquatic jungles; there were seasonal lakes and pools, just as in Doñana today, where the water table was close to the surface. These lakes attracted many kinds of ducks and other water birds whose remains we found inside the caves. There were also many frogs, toads, newts, and terrapins that would have come out to breed in the spring when water levels and warm temperatures made life an attractive proposition. Death would have come, like today, during the months of summer drought when life would have been hard pressed to continue unabated. Ironically, the Neanderthals would have had it worse not when it was cold but when it was hot and dry. Lack of fresh water would have severely limited their activities.

Not only did the Neanderthals have the grassy savannahs and seasonal wetlands outside their cave, they also had the tall cliffs of the Rock of Gibraltar where they could hunt ibexes, and then there was also the coast. For much of the last 100 thousand years of Neanderthal life on Gibraltar sea levels were much lower than they are now. This was because global temperatures were cooler than now and more water was trapped as ice in the poles. The coastal shelf off Gorham's Cave is very shallow and lowered sea levels of 80 to 120 metres, which would have been the norm when the Neanderthals lived there, exposed a huge area of land of what is now the sea bed of the Mediterranean Sea. The coast would have been up to five kilometres away from the cave, quite a contrast with today, with the sea lapping a small beach right by the caves. It is on this huge area, now under water, that the Neanderthals had their hunting territory. Many of the caves along this stretch of coast were inundated by the rising sea during the last global warming 10 thousand years ago; some are now completely submerged with all trace of the Neanderthals having been washed away. We are lucky that some of the caves, like Gorham's, were never reached by the sea and this has allowed us a glimpse into this lost world of the Mediterranean Neanderthals.

This lost world could best be described as a kind of Mediterranean Serengeti. Herds of grazing and browsing mammals roamed the grassy savannah. The main ones were red deer, wild boar, horse, aurochs, narrow-nosed rhino, and straight-tusked elephant; the latter two had become scarce towards the end of the time of the Neanderthals. This was a community of herbivores typical of the temperate forests and savannahs (Chapter 5). The array of potential food not only attracted the Neanderthals. Spotted hyaenas were the commonest meat-eaters throughout the 100-thousand-year period registered in the cave archive. The leopard, an efficient ambush hunter that relied on trees for cover and for hauling up its prey away from the hyaenas, was also here. There were lions, wolves, lynxes, brown bears, and a host of smaller carnivores too. So in this complicated landscape, the Neanderthals had to keep one eye on the prey and another on the predators.

The information that comes from these caves tells a very clear story of the way of life of these southern Neanderthals. They certainly ate large mammals and we know this because we find their remains, charred from being in the fire or with the telling cut marks made by the Neanderthals' flint knives as they cut through the flesh. Their main prey seems to have been the ibex; their remains outnumber the rest by far. Red deer seems to have been second on the menu then followed by the other grazers. This suggests that the Neanderthals may have been avoiding the dangerous animals, like wild boar, aurochs, and rhino, going for less formidable opponents instead. This is not surprising when we think that they were hunting by ambushing the prey with thrusting spears from close quarters.

We cannot be certain that they were hunting these animals all the time though. One feature of the seasonal environment of Gorham's Cave, like Doñana today, would have been the unpredictable droughts. Apart from the three months of no rainfall in the summer, some years the rains would not have arrived at all. Such years would have provided opportunities for scavenging. In Gorham's we find not just hyaenas but the remains of all four species of European vultures which show that a living was to be made from this activity. Once again the limiting factor would not have been food but water. The real surprise, though, came from the smaller animals found in the cave.

There is a well-established point of view among students of pre-history that denies Neanderthals and their contemporaries, excepting the Ancestors, the ability to hunt small game.[11] Birds, for example, were just too difficult to catch; paradoxically we saw in Chapter 2 how capuchin monkeys, without big brains and sophisticated technology, regularly caught birds in the rain forests of the New World. If they could do it then the early proto-humans probably could too, so why shouldn't the Neanderthals, with their large brains and tools, have been able to do so too?

A different argument does not deny the Neanderthals this ability but instead sees them as lazy or backward people who preferred to catch easy prey, only moving to more difficult animals when the stock of easy ones had been depleted.[12] In other words Neanderthals living in Mediterranean environments in Italy and the Middle East, where they had been studied, would spend their time beachcombing and collecting limpets and mussels from the seashore. The more they consumed the fewer were left and the evidence for this was claimed to be that the limpets and mussels got smaller as time went by.[13] So the Neanderthals, at a loss for what to do, set their eyes on tortoises. These were not static but they moved around slowly enough for the Neanderthals to be able to catch them.[14] When the tortoises were exhausted they turned to hares that were obviously much harder to catch and eventually they went for the really difficult animals—the birds. Unrealistic though this model is, it does deserve testing and we can do this with the information at our disposal in Gorham's Cave.

Gorham's Cave was truly a revelation for those who saw the Neanderthals as dumb and incapable brutes that somehow survived for over a quarter-of-a-million years on Planet Earth. For naturalists who have spent many hours in the field, Gorham's Cave simply confirmed the obvious—that prehistoric people, Neanderthals included, were resourceful, knew their environment well, and were not fazed by things that could run or fly. They were among the first, and probably the best, naturalists the world has ever known. So what does Gorham's Cave show us?

For a start, over 80% of the bones belonging to mammals eaten by the Neanderthals belonged to rabbits. These animals are endemic to the Iberian Peninsula and they were extremely abundant. There must have been thousands of rabbits outside the cave, where the sand dunes would have made ideal places for them to burrow. Catching rabbits obviously was not a complicated task for the Neanderthals and they ate a lot of them. One hundred and forty-five different bird species have been identified in these caves, making them the richest for fossil bird remains anywhere in Europe. The number of bird species found in these caves roughly represents a daunting quarter of all Europe's breeding birds. This is not surprising as the Strait of Gibraltar has been one of Europe's focal points for migratory birds commuting between African winter homes and European summer residences.[15] And the Neanderthals ate them.[16]

They also ate tortoises and the highly nutritious seeds of the stone pine,[17] and limpets and mussels too, just as in other coastal areas of the Mediterranean. The biggest surprises came when we discovered the remains of several Mediterranean monk seals that showed obvious traces of Neanderthal flint knives. The surprise became greater when the remains of fish and of two species of dolphin were found close by.[18] On the one hand it was a blow for those who advocated that the ability to exploit marine resources was the hallmark of the Ancestors and the unique behaviour that had allowed them to emerge from Africa by following the coast (Chapter 4). Here we had Neanderthals behaving in exactly the same way. More importantly, it showed that the Neanderthals that lived in Gorham's Cave, far from being specialized large mammal ambush hunters were actually general purpose foragers who could exploit the wealth of resources at their doorstep. The range was probably even greater than we had established but many kinds of potential foods, such as fruit, roots, and grubs, would have perished without trace. Large mammals were just a very small component of the total food intake.

So, in the south of Europe, the Neanderthals found a great diversity of resources and they seem to have been able to exploit them all. Further north, for example on the edge of the treeless plains of central Europe, large mammals were practically the only resource available to them—there were no limpets, tortoises, or seals here. There may have been freshwater fish in special places at specific times of the year and recent studies do not discard the possibility that these were consumed by Neanderthals. Why should that startle us anyway? Just watch brown bears catching salmon, without technology of any sort, and then ask yourselves why Neanderthals could not do the same or even better? All the same, these studies of Neanderthal diet in northern latitudes have concluded that they were strict meat-eaters,[19] hardly surprising given that would have been the main source of energy where they were living. That Neanderthals across the European plains, from Belgium to Croatia, appear to have largely eaten meat does not mean that they behaved in the same way elsewhere or at other times. This is yet another example of generalizing from limited information. Gorham's Cave clearly exposes the flaws of this particular extrapolation.

The next important lesson that this site gave us was that the range of foods exploited seemed not to have changed throughout the length of time that the Neanderthals had lived there. There had been no 'broad spectrum revolution', the kind of mussel-to-partridge evolution in diet that had been proposed for Italy and the Middle East. This was yet another revolution that wasn't. The Neanderthals had lived in the area for tens of thousands of years and had not overexploited their resources. The last Neanderthals of 28–24 thousand years ago were eating the same range of foods as their predecessors had been 100 thousand years earlier.

There was yet another crucial lesson to be learnt. The last Neanderthals were making stone tools and weapons no different from those of their ancestors. The last Neanderthals were still using Mousterian technology. Their world had not changed so there had been no need for them to change. Over 10 thousand years earlier, their relatives in France had moved towards new technologies, like the Châtelperronian, to be able to cope with the changing environments; all that time the southern Neanderthals stuck with the Mousterian. The same people had different cultures in distinct parts of their geographical range. Why should that surprise us today?

The world of Gorham's Cave and its surrounding area was a unique one. The region between Gibraltar and south-west Portugal behaved very differently from the rest of Eurasia. It was well south and well west; in fact some parts were further south than places in North Africa and the whole area was further west than Wales. This meant that its climate remained mild even during the worst Ice Age conditions further north. There were no high mountains that could generate a locally cold microclimate and the topography was varied, which allowed many species of plants and animals to survive in local populations in the many sheltered valleys of the region. The proximity to the Atlantic also buffered the area from extreme drought.[20] This little bubble, almost a piece of Africa in Europe as Chapman had described, allowed the last surviving Neanderthals to carry on with a way of life that had disappeared from the rest of their former range thousands of years earlier. Each time that the climate hit hard, there were fewer such pockets left until one day the Neanderthals were all gone.

Pockets of Neanderthals had managed to survive beyond the 30-thousand-year mark in places like the Crimea and the Caucasus but the suitable habitat available to them in these areas would have been much smaller than that in the Iberian Peninsula. Iberia, south and west of present-day Madrid, formed such an area. Much of it was lower ground than further north and it seems that the altitude change in this central plateau defined the southern limit for most of the steppe–tundra mammals.[21] This region would have been a major stronghold for Neanderthals. As climate deteriorated, parts of the interior of this region and the higher mountains would have become inhospitable and Neanderthals would only have survived in sheltered valleys.[22] The coastal areas would have remained as the main refuges. Those furthest away from the high coastal mountain ranges would have suffered least of all from the unpredictable climatic oscillations.[23] The Rock of Gibraltar, situated on a low-lying coastal plain well away from the major coastal mountains, was an ideal refuge.

The north of Iberia was a different world. It was part of the northern Eurasian system of mountains and plains that I described in the previous chapter. The southern slopes of the Pyrenees and the Cantabrian Mountains faced the tablelands of northern Spain that averaged 1000 metres in height. When it was cold and dry these tablelands would have formed a southern extension of the steppe–tundra of central Europe. Woolly mammoth, woolly rhino, reindeer, steppe bison and saiga antelope all penetrated into this area. The southern edge of the Pyrenees and Cantabrians were therefore a mirror image of the north. It is not in the least surprising to find Châtelperronian and Aurignacian technologies along with these steppe–tundra mammals here too. This was the south-western limit of this world and the presence of these animals and the Châtelperronian and the Aurignacian here but not further south confirms that fauna and technology were closely linked.

It was no coincidence then that the last Neanderthals survived in the extreme south-west of Eurasia. The Strait of Gibraltar seems to have prevented them from moving further south into North Africa—they had no get-out clause and they died out. Like the last populations of living pandas or tigers they became an endangered species with few chances of pulling through. In the end the last ones may have died out simply because so few of them were left that they were inbred, or a random fluctuation in numbers brought them to zero or perhaps a disease swept across the last population. We may never know the real reason of the demise of the last Neanderthals but we should be clear that this was not the cause of the Neanderthal extinction. As we have seen in this chapter and the previous two, the extinction was a long drawn out process of attrition over millennia.

There is something odd in the Gorham's Cave record that offers an insight into what may have happened to the last few. The Neanderthal camp fire that produced the charcoals that were radiocarbon dated to 28–24 thousand years ago was strategically located within the cave. Neanderthals normally placed their hearths at the entrance to their caves where the smoke would not fog up the cave itself. But Gorham's is a huge cave with a high ceiling which is over thirty metres above the cave floor. The Neanderthals placed their hearth deep in the cave without risk of suffocating. Behind the hearth was a large chamber where they could sleep protected by the fire that closed the entrance and kept predators out. They seem to have used the same spot over and over again as the earliest dates for the use of the hearth went back to 32 thousand years ago.

Another hearth was found close by, not more than a metre away, but this one had been made later, after 18.5 thousand years ago by Ancestors. Between one and the other, after 100 thousand years of occupation, was a long period when the cave was empty. Looking at climate records from marine cores taken from the Mediterranean seabed east of Gibraltar revealed a period of unusually harsh climate, marked by cold and drought, precisely when the cave had been abandoned.[24] For a while Africa in Europe had become a drought-stricken land, not severe enough to cause the extinction of plants and some of the more resilient animals but bad enough to tip a stressed population over the brink and prevent new ones from settling.[25]

The new people, when they eventually came, were the Ancestors and had a new, projectile, technology—the Solutrean. Here, at least, they were not responsible for anything that happened to the Neanderthals because the two never met. They painted deer and left hand imprints on the walls of the cave and they made necklaces from the polished and perforated teeth of red deer and from periwinkles. These were the differences between them and the Neanderthals. Otherwise they seem to have behaved in much the same way, catching the same kinds of animals in similar proportions. Location and available resources determined very much what people, Neanderthal or Ancestor, did here as indeed it did everywhere else.

The Solutreans arrived in Gorham's Cave from the north. This was the last place in Europe colonized by the Ancestors before the last Ice Age. As the cold conditions worsened many populations of Ancestors in central Europe went extinct and the south became a refuge once again. In the next chapter we will look at how these Ancestors spread across Europe and where it was that they came from. To find out we will return north and east into the steppes of Russia and Central Asia.

8

One Small Step for Man . . .

THE kinds of rapid environmental changes that Neanderthals had been experiencing across Eurasia, on either side of the mid-latitude mountainous belt, since at least 50 thousand years ago also affected the newly arrived Ancestors. We saw in Chapter 6 how difficult it was to distinguish the makers of the different technologies springing up here and there across this vast stretch of land; we were unable to confirm the presence of Ancestors anywhere north of these mountains before 36 thousand years ago. Even then there was some argument as to how 'modern' these people really were as they revealed features in their bodies that some anthropologists interpreted as indicating that they were mating freely with the Neanderthals.

It is after 30 thousand years ago that we find a clear and unequivocal signal of the Ancestors across the plains of Europe and beyond.[1] These people, descended from those that had entered the plains via Central Asia, were by then adept at making a living in an alien, treeless, landscape. They had managed to tear themselves away from the manacles that trees had imposed on all who had come before them and, indeed, on many others who were their contemporaries in other parts of the world. We could make a case for similar incursions away from woodland and savannah among other humans of common ancestry with these plains Eurasians, for example in Australia or parts of tropical Africa, but the divorce was most absolute and complete in the steppe–tundra of Eurasia.

The reason why the break was so unconditional among these people had to do with two features of the ecology of the northern lands. On the one hand the steppe–tundra was a haven of food for those who could find a way of exploiting it and on the other, water, so limiting in the tropics, would not have restricted movement to such a large degree. Here were large herds of reindeer, steppe bison, horse, and many of the large mammals that we have met in previous chapters. Nobody had ever managed to tap this resource except along the edge of this vast treeless landscape, where they could rely on some form of ambush hunting from within cover. These edge habitats also offered opportunities within the trees so people were never fully dependent on the unpredictable herds that came and went with the seasons. By living on the periphery of open plains and the woodlands they had lived in the most ecologically rich and profitable places.[2]

We saw in Chapter 6 how the constant changes in habitat that these edge people, probably Ancestors as well as Neanderthals, were exposed to, and the resulting pressure on their way of life, stretched their ingenuity. But the frequent and rapid changes to the landscape that took place with each climatic reversal brought them back to their old ways so there was never an opportunity to fully develop the newly acquired skills. A hypothetical example, probably quite close to what may have happened on many occasions, will clarify this point.

Imagine a group of humans faced with a landscape change, within two generations, which removed all the trees. There was still food out there: they could see tempting herds of reindeer on the open plains but they needed to get to them. One clever person in the group invented a light-weight spear that could be thrown from a distance. Now the clan members could gang up on an unsuspecting reindeer; but it needed many light-weight spears to bring the animal down so they plotted and planned on ways of chasing and isolating animals into the path of those waiting to throw the projectiles. It required group cooperation of a kind that they had never needed before.

Now the climate warmed up again and the trees returned. The reindeer had gone and the animals of the woods were red deer; these animals did not mass together in the numbers that the smaller reindeer did. The children of the clan hunters now found the light-weight spears useless. For one thing the animals were so scattered that chasing one took huge amounts of time and, in any case, those waiting could not see the animals coming with sufficient time to react. When they did they found that trees got in the way of the path of the spears. Even those spears that got through could not easily penetrate the hide of the large red deer.

One clan member remembered a hunting technique that he had heard from his grandfather who had lived in the old woods before they had vanished. He did not know exactly how his grandfather had done it but, through trial and error, he made a powerful new weapon: it was a thrusting spear with a lethal point made of flint which was hafted to the end of a long wooden shaft. Practice makes perfect, and the pressure of having to feed a starving family sharpened the mind. The now small clan found ways of ambushing red deer at close range, just like their ancestors had done. They ate and they survived. Not all groups made it though. Another clan in the next valley did not have a clever youth with a memory of stories told when he was a child. They all died that warm summer.

Further east, on the Russian Plain north of the Black and Caspian Seas, in Central Asia and in southern Siberia, people would have been exposed to treeless habitats for much longer. The climate was always drier and much more continental than in the European Peninsula, the Far West.[3] Here the changes would have been largely dictated by rainfall, shifting steppe to grassland to open savannah and back. At the worst times, deserts would have taken over some parts, much as they do today over large areas of Central Asia. By and large though, the major habitats were treeless with woodland and savannah occupying the foothills and slopes of the mountains to the south. So here, in the centre of the Eurasian land mass, people were also able to live on the rich edge between wooded and treeless habitats; but the tantalizingly bountiful steppe, unlike Europe further west, was ever present here and people had the opportunity of developing ways of exploiting it without having to keep returning to old habits when climate changed.

We have practically nothing by way of human remains to guide us through this period in this vast region but genetics tells us that a branch of the Ancestors that had reached India and nearby regions spread northwards, possibly exploiting corridors of steppe habitat, and broke through into Central Asia some time between 50 and 40 thousand years ago.[4] Other populations may have entered Europe, as we saw in Chapter 6, but gained only tenuous footholds and left little genetic evidence of their temporary sojourn there.[5] Technologies defined as belonging to the Upper Palaeolithic begin to show up across the Russian Plain, Central Asia, and southern Siberia after 45 thousand years ago and, despite the absence of human remains, have usually been attributed to the newly arrived people.[6] This upsurge of novel technology was part of the widespread changes affecting the whole of Eurasia after 50 thousand years ago as climate became increasingly unstable (Chapter 6).

Old and new habits mixed in the world between steppe and woodland. In these areas some cultures seem to have emerged from an existing local base, the Mousterian of the Neanderthals, while others appeared as sudden ruptures which have been taken to mean the arrival of new people.[7] Upper Palaeolithic and Middle Palaeolithic cultures overlapped across this vast space for eighteen millennia; there is no better signal that changes were gradual and did not affect all populations equally. The problem, as was the case in central Europe, is identifying who was making what. Overall we must conclude that, as in Europe and the Middle East, there was no simple takeover of the Ancestors on arrival.

From these humble beginnings on the edge of the steppe around 45 thousand years ago we find a thriving population of undisputed Ancestors out on the plains 15 thousand years later. Neanderthals are to be found nowhere in this scene. The remnants were confined to refuges further south (Chapter 7). Difficult as the threads are to disentangle, the outcome of this long period of constant attrition of human populations caused by rapid climate change is the exclusive presence of an ever-expanding population of Ancestors on the Eurasian Plain. This plain was only marginally exploited by earlier peoples, including Neanderthals, so its invasion was not at the expense of others. People had broken through a new frontier where no one had been before. It truly was a giant leap for humankind.

The story of how some humans managed to make this breakthrough was most likely one of trial and error with many failures along the way that we will probably never be able to pick up from the archaeology. As with much of written history, archaeology is most likely to reveal to us the success stories and not the failures. The reason is simple enough: the successful peoples would have been those that produced the largest populations which in turn would have been most likely to leave their imprint on the soil. One thing that we can pick up with some degree of confidence is the point of origin of this breakthrough and it is the genes of present-day people that give us the tell-tale signals.

Central Asia shines through as a region of huge importance as a reservoir of genetic diversity; they had more time to accumulate genetic novelties here than anywhere else in Eurasia and it is from here that people expanded across the plains, westwards towards Europe, and eastwards towards the Pacific coast of Asia and, eventually, into North America.[8] The source Central Asian population carried a unique genetic marker that originated some time between 40 and 35 thousand years ago. The people who spread westwards into Europe carried another identifier genetic mutation, one derived from this parent stock and which indicates a westward expansion around 30 thousand years ago.[9] So in this case, at least, the genes seem to match the archaeology and the human fossil evidence.

The birth of Europeans, northern Asians, and Native Americans was here, in Central Asia, between 40 and 30 thousand years ago. Their success and expansion arose from a combination familiar to us by now: a people had found themselves in a place where herds of grazing mammals roamed freely across open steppes. Here the steppes changed little from year to year and these humans found ways of exploiting its animals. There was nothing particularly special or unique about these people—they were simply lucky to have found a place that nobody had been to before. They thrived. As climate got colder and drier this homeland started to become desert but the steppe and its animals moved westwards taking over lands that had been the domain of the trees. People followed, just as the collared doves in our Prologue did, and they too eventually reached all the way to the Atlantic coast of Europe. Others went east but we will leave them until the next chapter.

In his book *Guns, Germs and Steel*, Jared Diamond showed how geography and ecology gave some people a technological edge over others after 10,000 years ago when the climate of the planet had settled down to something similar to today.[10] The cultural and technological differences between people had developed because of location and circumstances and had nothing to do with having superior brain power or abilities. As we have seen so far in this book, the argument can be extended much deeper in time and probably even applies to situations which affected populations of humans that were more different from each other (e.g. Neanderthals, proto-Ancestors, and Ancestors) than was any post-10,000-year human population. Like the big revolutions in agriculture, farming, and industry, these earlier leaps had much more to do with circumstances than with any kind of intellectual differential.

The best known milestones of our history fed themselves and grew in an autocatalytic manner—once started there was no going back. That was the case in practice at least, but in theory it need not have necessarily gone this way. Because the climate after 10,000 years remained mild compared to what preceded it, agriculture, farming, and human populations were able to grow exponentially in many parts of the world at the expense of hunting and gathering existences. Had climate deteriorated once more, these technologies may have become defunct like so many others did in previous times. The way in which our population has grown since the discovery of agriculture has been vertiginous within a tiny interval of time. Between 45 and 30 thousand years ago, Eurasian humans experimented at exploiting the cornucopia of the steppe–tundra but the wild oscillations of the climate kept pegging them back. Then the breakthrough came for one population somewhere in Central Asia. The discovery of agriculture, some 10 thousand years ago, is usually hailed as the starting point of our special history— as we turn now and look at the achievements of the plains people of 30 thousand years ago, we will realize that it was there and then that it all started and that the people of post-Ice Age Eurasia simply adopted and adapted what others had already invented thousands of years earlier.

How do you deal with a flat and endless landscape where there are no trees from which you can stalk your prey or escape to if chased by a carnivore, no caves which you can shelter in, and no features that can prevent you from getting lost? It all looks the same and going into it can appear a scary and daunting proposition. Every so often, from the comfort of a cave up in the hills along the edge of the plains you see a herd of juicy reindeer, waiting to be taken, and the temptation becomes too great and the opportunity too good to be missed. You probably start off with forays, maybe at particular times of the year when the animals are most abundant or most vulnerable, but you never lose sight of the broken relief that is home. You invent new weapons and techniques that allow you to hunt in the open but you do not abandon the old ways completely. You still ambush the odd ibex or red deer back in the hills. They are your life insurance and you have become a sophisticated risk manager.

You have developed a mixed strategy, one of many similar developments that neighbours and distant cousins have been practising for generations. Sometimes you meet these neighbours, or you spy on them, and you pick up new ideas; they do the same to you when you are not watching. Sometimes your neighbours are friendly and keen to cooperate, share technology, or exchange commodities.

Other times strange people who speak in an odd tongue enter the area but you are unsure of their motives. They are powerful and muscular so you keep a safe distance. They soon move on, apparently keener on the solitary red deer stags dispersed all over the hills than in the big herds of reindeer down below. You have seen them trying to catch the reindeer before and you have been impressed by their weapons, some of which you have copied, but they are awkward out in the open and they lose more animals than they get despite their inventiveness. As they pass, your attention turns once more to the hundreds of tiny specks on the flat lands down below.

Threads of smoke rise over a collection of strange domed structures crowded together close to a river meandering in an otherwise featureless landscape. We have travelled in an imaginary time machine back to 28 thousand years ago and a remote place somewhere north of the Black Sea. As dawn breaks wolves howl and bark, surprisingly from the area where the structures are situated. A man comes out from one of the domes that appear to have been made from animal hides. He carries a large piece of meat which he throws to the odd-looking wolves now quite visibly tethered. Noisy children emerge followed by women, probably their mothers. Soon the camp has come to life and people of all ages and both sexes are performing different activities. Some talk to each other while others just get on with the tasks. The children watch and play. The whole camp appears to behave like a giant superorganism, like a nest of ants. But these are not automated giant ants, they are people, and their daily routines are interspersed by large doses of improvisation and activities that have clearly been shaped by their experience of what has gone on the day before. We have stumbled across a community of Ancestors who will become known by their cultural achievements and, even though they started their career in the east, their culture will become known as the Gravettian from a French site where it was first recognized.

In 1908, during excavation works in the Willendorf area, related to the building of a railway line between Krems and Grein in lower Austria, a worker discovered a small female statue that became world famous. It was the Venus of Willendorf, a magnificent example of the kind of art that the Ancestors were making on the European Plain around 26 thousand years ago. Right across the plains of the major eastern European rivers, such as the Danube and the Don, Ancestors with a new culture—the Gravettian—were settling in large camps that could best be described as hunters' villages. In the east, on the Don River, Russian scientist I. S. Polyakov had been excavating such villages since 1879, in Kostenki. Sites that have become the hallmark of this culture—Dolní Věstonice, Pavlov, Avdeevo, and Kostenki itself—include the remains of huts, kilns, storage pits, tools, jewellery, and figurines.

The Gravettian people were the hallmark of human cultural and technological achievement. Their origins were in the plains north of the Black Sea or even further east, and their genetic roots go back to those Central Asian people who first learnt to tame the open steppe. With their skills and ingenuity the Gravettians managed to succeed where others had failed and they achieved this at a time when the climate was building up to the height of the last Ice Age. From 30 until 22 thousand years ago, almost as long as the later world of farmers, the Gravettians ruled the roost across the western Eurasian Plain and they managed to penetrate south into the Mediterranean Peninsulas, with pioneers even reaching the south-western extreme of Iberia. They probably had little contact with the Neanderthals—by the time they reached the last outposts of these people, the Neanderthals were long gone.

So, despite their technological and social achievements we cannot hold the Gravettians responsible for the demise of the Neanderthals. Since they would have rarely met, it comes as no surprise that all present-day descendants of these plains people, who are of European and western Asian ancestry, show no trace of Neanderthal genes whatsoever. Even less contact would have taken place with the branch that went towards eastern Asia and North America, and less still with southern Asians, Australians, and Africans who never came into contact with the Neanderthals. We now need to look at what the archaeological record has left of the prowess of the Gravettians to understand how they were able to succeed where nobody had been able to before them.

There was little new about the various elements of Gravettian behaviour on the plains. In fact much of what has come to be considered modern behaviour, and part of the 'revolution' that made us, had already been invented tens of millennia earlier.[11] The real novelty lay in the combination of these into a single package and the cause of this amalgam was the need to survive in the treeless landscapes of the Eurasian steppe–tundra. We picked up some of the components of this package when we looked at the way in which Neanderthals, and presumably pioneering Ancestors, tried to cope with this new landscape across Eurasia between 50 and 30 thousand years ago; the appearance of the package on the plains with the Gravettian people 30 thousand years ago would have been the result of thousands of years of experimentation and information transfer between hunters from different parts of this world.

Recall that all of the behavioural novelties of the Gravettian culture can be understood without recourse to explanations that involve sudden changes in brain wiring which produced particularly brainy people. These people were already smart and so were many of their contemporaries, and predecessors, many of whom never made it. The innovations had instead to do with the pressures imposed by the new terrain in which they were living. Imagine having to live in a desolate, treeless landscape in which getting your bearings to avoid being lost or finding the large herds of grazers was as complicated as navigating in the open ocean. This was also a strongly seasonal world, one in which a new experience—the long nights during the cold winter months—had to be dealt with somehow. The biological make-up of these people, descended from a long line of tropical primates, did not make hibernation an option. Naturally people had to find ways of moving around without having to carry heavy loads; they also had to find ways of reading the land and communicating with each other with precision. The Gravettians had entered the information age.[12]

Many of the inventions that have been attributed to being modern can be simply understood as ways of dealing with this environment. Take the use of bone, ivory, and antler to make tools and weapons. These materials are quite typical of Gravettian sites.[13] Major sources of antler were reindeer while mammoths contributed ivory and bone.[14] These products would have been readily available to people living on the plains but not to those living in other environments without such animals. Their use reflects availability and also ingenuity on the part of people who could turn these into functional, and in some places also decorative, objects. For much of human history, people would have used wood as the prime raw material for tools and weapons. Rarely do we find evidence of such wooden objects as they perish more readily than stone but the available examples show that wood was already being used by *Homo heidelbergensis* (Chapter 5). People finding themselves in an environment in which trees were a scarce commodity would soon have turned their attention to other organic materials, like bone, antler, and ivory, as substitutes.

The places that the Gravettians and their kin frequented would have been cold for much of the year with the ground frozen for long periods. This frozen ground, the permafrost soil, would have acted as a kind of natural freezer that slowed down the decay of dead animals. As these rotted slowly, or were exposed by the thawing permafrost in summer, they would have become available to people. There must have been lots of carcasses and skeletons lying about in places and people soon found uses for these, particularly the remains of the mammoths. In the absence of natural rock shelters, the people of the plains had to construct their own and the large bones and tusks of the mammoth made excellent superstructures that could be covered with animal hides to make tents. Some of these structures were partly dug into the ground and similar ones would have been made earlier, using wood, by people first venturing onto the plains,[15] and probably by many others, including the Neanderthals, who had the capability of working wood.[16] Tent building is a good example of a technology that probably developed gradually but which suddenly received a big boost as people came under pressure to find shelter from the icy winds of the plains. With a shortage of wood, but an abundance of large bones, tusks, and hides, architecture leapt forward.

The Gravettians seemed to focus their attention mainly on reindeer, also horse and hare; scavenged mammoth remains were brought back to camps and the inedible parts were gradually accumulated. Sometimes, it seems that Gravettian people may have taken the initiative and got together in large groups at known locations where mammoths passed or gathered. Here they took advantage of local features of the landscape, such as narrow valleys and marshy ground, and organized collective mammoth hunts. These sites are rare in comparison to those with reindeer and they seem to be examples of extreme solutions in situations of severe stress.[17]

These organized hunts perhaps took place when reindeer were scarce. Hunting the large mammoths would have been a daunting proposition and could not have been achieved without close cooperation among hunters. Facing a charging bull African elephant today, even with a rifle, is a dangerous and unpredictable affair. Challenging a large mammoth covered in thick hair that was an effective shield against the Gravettians' spears would have needed a lot of ingenuity, courage, and cooperation as well as an intimate knowledge of the terrain. We can only guess at how they did it. Perhaps they isolated weak or young animals, or they possibly used fire to cause stampedes towards natural or prepared traps; it seems unlikely that they would have faced them head-on with spears. One thing seems certain: they would not have attempted such a feat alone or in small groups. Cooperation was the key to success.

The permanence and size of some of the Gravettian camps does suggest that these people had taken the unprecedented step of amalgamating small bands of hunter-gatherers into larger collectives. Perhaps this was happening only seasonally or in particular sites but it marked the beginning of a new world, one in which humans were able to adopt new ways of life simply by cooperating with each other in numbers. At the heart of this new way of living was the base camp, an early kind of village; what was new about this when humans had been gathering around camp fires for hundreds of thousand of years? The difference had to do with the environment they were in. This environment was a vast expanse of emptiness. If you found a herd of grazers you had hit the jackpot, but what did you do in the meantime?

There are many advantages to gathering and cooperating in large groups but these benefits do not come readily in some situations. The ambush hunters that went for isolated animals or small herds dispersed in woodland or savannah could not have hunted the quantities of animals needed to sustain a large population. Large groups of people were simply not viable in those circumstances. In situations in which larger herds might have become available, for example the cases of the seasonal concentrations of grazers on the tropical African plains, the absence of ways of preserving the meat in hot climates and of carrying it back in large quantities to base camp also made large hunting groups impractical.

In the new world on the Eurasian Plain, the cold climate allowed food to be preserved for much longer than in the tropics. The Gravettians dug pits into the ground and were able to store food in natural freezers.[18] They had invented an economy in which surplus catered for risk. Storage then became a central element of this way of life. With it came the need to protect and curate the stored supplies. For parts of the year at least, those times when foods were kept within the Gravettian village, some people in the group would have been confined to its immediate vicinity. We can conclude two things from this observation: on the one hand some people's nomadic hunter-gatherer ways were curtailed and, on the other, there must have been a division of labour within the group. Quite possibly specialists at different crafts arose from this early separation of tasks among the group members.

It is hard to think of a Gravettian hunting clan as being totally sedentary. Groups would have had to make regular forays to catch the animals that were the mainstay of their economy. Perhaps these people were seasonally nomadic, moving camps in accordance with the disposition of resources, much like the Mongol nomads do even today. The dark winters may well have been spent close to the villages, surviving on the stored foods and away from the dangers of packs of marauding wolves. Spring, summer, and autumn would have been the main hunting seasons and villages may well have been positioned close to the expected paths of herds. Even so hunters would have had to range widely, away from their base, to maximize their efficiency. The autumn hunts would have been particularly important as animals such as reindeer would have been fattest, and have had the thickest coats, then. The meat would have preserved best in the cold and dry autumn air.

The larger the group, the more mouths there would have been to feed and the more surplus would have been needed for the bad times. During the hunting season the village probably acted as an information centre, much as colonies function in birds.[19] Imagine a newly arrived clan that has set up a village for the autumn hunting season. Once established, decisions must be taken regarding the possible location of the herds. The experienced hunters discuss the likely options but, as they stare across the expanse of flatness, the decision is taken to split into small reconnaissance parties and move out in different directions. The distances are vast. A herd might be encountered after an hour or it could be days before the first animals are sighted. The forays are planned and a time agreed to meet back in the village.

Three days later the different groups return and report on their sightings. Some groups saw nothing or just scattered animals. Two groups found large herds and the planning of the hunt starts. The village has acted as a focal point and a place where information has been focused—it has become the community's nerve centre. Something fundamentally new has taken place in the process. It is something that may have happened sporadically to other hunters on the African and Australian plains but here it will leave an indelible mark. For the first time in human history some people experienced things that they had not seen for themselves. Some saw herds of reindeer that others did not and they communicated what they saw and where they saw it. The pressure now was for efficient communication and sophisticated ways of transmitting this information were required. The predecessors of these people, as far as we can tell, could speak—once again this was not a new thing. What was new was the need to make this communication by spoken language as detailed as possible. I have no doubt that the Gravettians must have had specialized terms and words related to their hunting first and foremost, and that the need for such terms must have driven the development of complex language. With it came other things intimately human: the wrong transmission or reception of information—error and deception.

Communication went beyond the spoken word. It may have translated into music, something difficult for us to confirm from the archaeological record. But we do have tangible figurative art, remnants of what has been seen as a cultural zenith and hallmark of modern humanity, even though this too was not that special. There is every indication that people controlled ochre, other media, and created art as far back as 160 thousand years ago.[20] There are claims for rock art going back to the Aurignacians in Europe, though we cannot be absolutely certain that these were not actually done by Gravettians.[21] The big surge in art came with the Gravettian people.

On the plains it took the form of portable art, small statuettes and sculptures often made out of clay baked at high temperatures in kilns.[22] The origin of pottery is usually attributed to the Middle East, around 8 thousand years ago, but the earliest clay pots date back to at least 13 thousand years ago in eastern Russia.[23] Although they did not make pottery vessels, the Gravettians of the Eurasian Plains had mastered the fire-using technology needed for ceramics 15 thousand years earlier than the first potters of the Russian Far East and 20 thousand years before those of the Middle East. We pick up other evidence in their paintings in caves in south-west France, in sites like Pech-Merle and Les Garennes. These paintings are of perfect execution which suggests that the Gravettians had been painting for a long time prior to this, presumably in perishable media in the open, and that the preservation inside caves simply retained a biased sample, reflecting the numerous limestone caves in south-western France, for us to marvel at.

The significance of cave art has been given disproportionate importance in the study of prehistory. This is, in part, understandable as much of it is very attractive and beautifully executed but the number of caves with art, and their geographical location, clearly indicates that the phenomenon was not a widespread one and cannot be used as a milestone of humanity's progress. The available evidence favours instead a gradual, patchy, and localized appearance, disappearance, and reappearance of art rather than a sudden and singular revolutionary affair.[24] It is also important to bear in mind that not everyone painted in the caves of south-western France or north-western Spain; there were masters that can be identified by their style and execution,[25] but most people would have been incapable of executing such wonderful images just as few of us could hope to replicate the Sistine Chapel. And while some of this wonderful art was made by the earliest Ancestors to reach Europe, for example around 30–27 thousand years ago in Chauvet and Les Garennes in France, most of the best known sites (Lascaux in France, Altamira in Spain: 17–14 thousand years ago) are thousands of years later, after the end of the last Ice Age, and long after the Gravettian culture.

We may never know why some people took so much trouble to paint such beautiful images deep inside the caves of a region of western Europe. I am of the opinion that the origin of such art had to be functional and the preponderance of animals, especially the grazing herbivores that were potential food, in the cave panels may have had to do with the transfer of information about hunting. We may never know the real reason that drove these masters to paint where they did. That the animals depicted are almost exclusively those of the open steppe–tundra closely implicates the Gravettians in the origins of this tradition, one that we often forget was subsequently lost by the hunter-gatherers of Mesolithic Europe.[26] And is it not surprising that nobody seems to have ever depicted a Neanderthal on the walls of a cave? Perhaps it is not surprising if the painters had never come across a Neanderthal.

Until the Gravettians had found a permanent way of successfully exploiting the steppe–tundra, the main factor that limited the movement of people would have been water especially in the tropics but also in other semi-arid regions, for example the Middle East and the Mediterranean lands. The reason that the Gravettians would have been less restricted was that water would not have been a real problem for them. The steppe–tundra was a land of lakes and there was also a huge reservoir of frozen water trapped in the permafrost.[27] The Gravettians knew about this as they dug into it to make storage pits so they must have known how to melt ice back to water with the use of fire should the need have arisen. Water was no longer an impediment to mobility and would have been a much needed resource for people relying heavily on meat for food and having to dispose of unusually high loads of urea as a result.[28]

We have seen how shortages of wood were overcome by resorting to bone and other organic materials but what about stone? Human technology has been inexorably linked to stone tools from the earliest times, perhaps disproportionately so because stone implements have survived in much better state and quantities than wooden ones. People venturing onto open plains would have found themselves away from the quarries of flint and other rocks suitable for making tools and weapons. The solution was to either make do without stone implements or to carry them in their hunting trips. To do the latter needed a change of mindset. Carrying large quantities of heavy stones would have slowed these hunters down so the option, which we began to see among the Aurignacians and Châtelperronians, was to make smaller, light-weight armatures and points that could be reused over and over again.

Economical ways of getting the most out of flint nodules that might be found scattered here and there on the route would have also been an advantage. The manufacture of tools and weapons from prismatic blades that were struck from the flint nodules became the fashion. This technique involved the preparation of flint cores into polyhedral shapes from which long, narrow blades with parallel edges were struck off using a hammer. By making these blades, incidentally not an exclusive invention of the Gravettians, instead of the traditional wide flakes, people could get much more out of a chunk of flint and depended much less on having to keep going back to the quarries in the hills. Such new methods of getting the most out of a piece of flint went well with the new projectiles that were being developed and that allowed people to hunt animals from some distance. The double advantage of portability and hunting efficiency tipped the balance in favour of blade-making technology. We will see in the next chapter that the trend towards making smaller and smaller weapons continued well after the Gravettian culture had gone.

For a long time it had been assumed that technologies involving the use of plant fibres were a late invention, one that came with the development of agriculture. Then the results of a series of stunning and imaginative studies were published by James Adovasio and Olga Soffer at the University of Illinois; they looked at the detail of the impressions

Figure 12 Excavation during 2007 in Pavlov VI (just adjacent to Dolní Věstonice, Czech Republic) showing a central hearth, round pits and mammoth bones

Source: Photo credit: Dr Jiri Svoboda

left in the various ceramic statuettes and other artefacts made by the Gravettians. A detailed examination of the Dolni Věstonice and Pavlov artefacts from the Czech Republic revealed that plant fibres were in use to make textiles, basketry, cordage, and probably netting between 26 and 25 thousand years ago (Figure 12).[29] These results showed that this kind of textile and basketry work had started in this part of the Eurasian Plain between 7 and 10 thousand years before anywhere else, not dissimilar to the start of ceramics. The results coincided well with known archaeological firsts of the Gravettian, for example the invention of eyed needles made from bone which must have been used to sow hides and textiles.

Although it was not possible to prove beyond doubt it seemed that these people had also found ways of making fine-meshed nets that would have been used to catch hares and foxes in particular, animals common in Gravettian archaeological sites. The foxes were presumed to have been hunted for their furs and the hares both for their furs and as a source of food. The implication of these results was that the Gravettians were among the first to practice mass harvesting of small mammals by using nets. You no longer needed to be a powerful and muscular hunter to have a regular source of sustenance.

Physical build is at the root of the success of the Gravettian story. These people were much less robust and bulky than the Neanderthals and some of the early humans who also practised close-quarter ambush hunting. Whether the Gravettians inherited this build from African ancestors or whether they gradually evolved it is a matter of conjecture. I favour a combination of the two with a gradual refinement of body form in a people constantly exposed to the plains of Central Asia and its animals for millennia. In the same way that weapons had to be lightweight, any change in physique that favoured energy efficient movement over the plains had to be favoured by natural selection. Other people further west, Neanderthals and Ancestors, were not exposed long enough to these environments and so could not adapt.

Genetic continuity between Neanderthal populations, most of which were in the hills, meant that any genetic novelty that made some individuals more lightweight would have been quickly swamped out since it held no advantage. In the case of a population of Ancestors, already with a gracile body form and geographically isolated from others of its kind, and genetically isolated from the Neanderthals, the lightweight body plan was rapidly favoured.

As the climate changes became frequent and the steppe–tundra encroached, humans of all kinds and descriptions attempted to deal with change by developing new technologies. In the end their bodies let them down and the best technology could not make up for the deficit of having to cover huge areas of ground in search of animals. The Neanderthal extinction was the extinction of a particular body form that had been around for a long time.[30] It was a body form not exclusive to the Neanderthal but to many other humans too. That one population of Ancestors was able to change its physique in one part of the world where it was continuously exposed to open environments was the result of circumstances. That the environment to which this body form was suited should have subsequently expanded so greatly was again a matter of pure chance. The jump onto the steppe–tundra was indeed a small step for a population of Ancestors, but its impact was to be greater than Man's first landing on the Moon, thirty millennia later.

9

Forever Opportunists

THE spreading out of a Central Asian population into Europe around 30 thousand years ago was only part of a wider picture. There is no reason to suppose that such an expansion followed a single direction. After all this was not a concerted drive or migration, like the raids of Genghis Khan and his Mongol hordes would be many millennia later, but a demographic collared dove-style expansion. We have seen the driving force behind the expansion and the human response is what biologists would call ecological release: a population adapts to a new, untapped, environment and spreads rapidly in the absence of competitors. There are many such examples of animals that have been introduced into foreign lands by people, often with hugely negative effects on the local endemic fauna.

The rapid expansion, practically reaching France from the steppe within a millennium,[1] may have been the result of a simple increase in birth rate and decrease in mortality as a result of improved resources.

The catalyst of the rapid expansion may have had to do with the new way of life of the plains people. In the previous chapter we touched upon the village or camp site as an information centre, allowing different groups of hunters to exchange data about where the herds were located. The efficiency of hunters on the open plains would have been seriously compromised if all the individuals of the group—old, young, pregnant women, weak individuals—went along. Such movement of entire kin groups might have been possible in other environments where territories were smaller, and it was probably the way the Neanderthals moved about.[2] But over the large distances that would need to be covered in a short space of time on the plains these non-hunter individuals would only have slowed down progress and got in the way.

At the same time these people might be usefully employed back in the village, performing other tasks just as important as hunting. Such tasks might have included, for example, curing meat and storing it for future use, basketry, textiles, or making and firing ceramics. None of these activities would have suited groups of nomadic hunter-gatherers in which all members moved. So something as simple as entering the plains may have forced some Ancestors to develop the semi-nomadic lifestyle that we find among the Gravettians.

A consequence of all this, and the one that might have inadvertently held the key to the ultimate success of these people, would have been a reduction of the interval of time between child births in women. In the fully nomadic lifestyle of the Neanderthals and many groups of Ancestors, families would have needed to raise children to some level of independence before the next pregnancy. Such a strategy may have put a premium on precocity among children, something which appears well established for Neanderthals. Neanderthal children appear to have had, on average, an accelerated rate of development relative to Ancestors' children: the dental development of a Neanderthal child aged eight at death was, for example, comparable to that of modern human children who are several years older.[3]

With a part of the community staying behind in the village, for at least part of the year, a woman would have been able to repeat pregnancy before her earlier child was fully independent. There would be less need for her to move long distances and there would have been other women, men, and grandparents about to help with the rearing. This cultural change would be expected to have subsequently led to positive selection of physiological attributes that optimized the new behaviour. The pressure on child precocity was also released. This simple shift in strategy of landscape use would have catapulted a rapid growth in the population.

The various elements that we observe as a package in the plains people (lightweight tools kits, projectile technology, mammoth bone huts, storage pits, portable art, use of fire technology for baking clay, base camps, division of labour, etc.) would have been possible only with this semi-nomadic strategy that would have allowed a division of labour within the group, time to make ceramics, decorate objects, paint, store food, make nets, clothes, and baskets, even for specialized craftsmen to make weapons and tools. It could also have predisposed people towards agriculture; after all growing plants would only have been possible if people stayed or returned to the area where the seeds had been planted. The reason why agriculture did not appear in the plains 20 thousand years before it did further south would have had to do with the kinds of plants available to the Gravettians, a harsh climate that would have been unsuitable for growing plants, and the semi-frozen soil.

Domestication of animals would have been another matter altogether, although there is no evidence that animals were kept and modified genetically. The reason may well have been that the kinds of animals that would have been potentially domesticated, except the horse, would not have lived in the steppes. We can only speculate whether horses and reindeer might have been herded at least. Such a strategy would have kept fresh meat close to the villages throughout the year but it is difficult to detect such patterns of behaviour from the archaeology, especially among animals that would not have been significantly altered from the wild types. But there is one candidate for early domestication that I alluded to in the previous chapter—the dog.

In Chapter 6 we saw how the wolf had become the ultimate long-distance pursuit carnivore of the open landscapes of Eurasia. If any other predators were able to challenge it in this terrain it would have been the Ancestors. Similar pressures for dealing with grazing animals in treeless landscapes generated similar results in animals of very different origins. Anatomy and behaviour predisposed these unrelated mammals to becoming the super-predators of the steppe–tundra. Humans and wolves became endurance runners that hunted in packs and sooner or later they were bound to come across each other.[4] Competition was one possible result; mutual cooperation, a kind of symbiosis, was another. The world of wolves and humans came together in these plains.[5]

Nobody knows when wolves were first domesticated. The first clear evidence of dogs comes from the site of Eliseevichi I in the basin of the River Dniepr on the steppes of present-day Ukraine.[6] Two skulls of dogs that resembled Siberian huskies were excavated along with remains of woolly mammoth, arctic fox, and reindeer and date back to between 17 and 13 thousand years ago. Though significantly more recent than the 30-thousand-year-old Gravettians, these modified wolves were clearly present in a similar ecological context. If the skulls from this date appear with clear signs of modification it must mean that the transformation of wolf to dog started, here at least, well before people settled down to a farming existence after 10 thousand years ago and the dog was therefore the first animal to have been domesticated by the Ancestors.

My hunch is that the intimate relationship between Ancestors and wolves started on the steppe–tundra and probably with the Gravettians and their contemporaries. Since the first tamed wolves would have looked just like their wild cousins any chance of picking up the trail from fossils must remain remote. Can genes tell us anything? They can. A study of mitochondrial DNA from 162 wolves from 27 different localities and 140 domestic dogs of 67 breeds came to the startling conclusion that the earliest dogs could have originated as far back as 135 thousand years ago.[7] If correct, that would put dog origins back to the time of the Neanderthals but even if the date was an overestimate it does support the idea of early domestication during the Ice Ages. The dog would have become an indispensible partner of humans, its social conduct allowing a close bond to be formed with people. The dog then became the first living tool and hunting weapon that humans used,[8] certainly a behavioural milestone thousands of years ahead of farming and agriculture. It was probably used additionally in the defence of the village from other predators, including people.

Not all encounters with other groups of humans need have been hostile. The evidence from the Gravettians and their contemporaries suggests that trade may have had an important role in their lives.[9] Certainly there is good evidence of long-range transport of stone, shells, and other materials from distant sources that may have found their way across the landscape through exchange. Such a system of trading networks would have served to buffer people even further from risks associated with unpredictable food supplies and they would have functioned particularly well across the almost barrier-free landscape of the plains. We are witnessing here the origins of open economies that stretched beyond the local needs of groups. Like the trading and related risk management systems of the aborigines of the Australian plains,[10] these networks may have operated in a clan-like manner that may have involved alliances that functioned over great distances and periods of time. Surplus supplies need not all have been stored: some may have formed part of the trading networks. Art and ornaments would have served as signals of identity among people and mobile art may well have also served as currency in trading deals. Perhaps it is not surprising then to find that the production of figurines in places like Dolní Věstonice (Chapter 8) reached industrial proportions.

With some or all of the different strategies that the Ancestors had packaged to be able to exploit the solitude of the steppe–tundra wilds, all of which were clear markers of Ancestor behaviour of a kind that we would certainly identify with, people spread across the landscape. They reached Europe in the west but they also ranged eastwards, north of the mountain barriers of the Himalayas and associated mountain chains, across Siberia and northwards to the Arctic which was reached as early as 36 thousand years ago.[11] As in Europe these may have been pioneering incursions ahead of the big wave of advance. If we follow the rate of spread that we calculated at the start of this chapter, Ancestors departing from Central Asia around 30 thousand years ago ought to have been in the Lake Baikal region shortly after (30-29 thousand) and in the region of the Bering Strait (the submerged continent of Beringia) by 28 thousand years ago. This is, in fact, what we observe and the genetic signal shows them to be of common stock with the ancestors of the Gravettians.[12]

Early eastward pioneering forays may have been of a similar nature to the pre-Gravettian arrival in Europe and these people also seem, like those in Romania and the Czech Republic (Chapter 6), to retain archaic features in their bodies. The best example of the few available is the individual from Tianyuan Cave in China that was found in 2003 and which has been dated to around 34–35 thousand years ago.[13] These early, pre-30 thousand, European and Asian human fossils all seem to retain traces of archaic features that have been interpreted as evidence of Ancestors hybridizing with archaic people (e.g. Neanderthals) as they spread. All these fossils come from within the confines of the mid-latitude belt of hilly and mountainous terrain which suggests to me that they were early pioneers from the Indian source area that had not quite mastered the treeless landscapes and were staying on the edge habitats. They may have hybridized with the people that they met but this is not required to have happened to be able to explain the archaic features. They were a northern branch of the more robust Ancestors that headed for south-east Asia and Australia that had not lost all these features. It is from one such population somewhere between Europe and China that the gracile plains people emerged.[14]

If we leave the people of the steppe–tundra for a moment and return south of the Himalayas we find that other branches of the Ancestors that had been spreading eastwards towards Niah and Australia (Chapter 4) were trickling northwards towards inland China via river systems, but a more important expansion was following the coast northwards towards Korea and Japan.[15] We saw in Chapter 4 that the pressure to stay along rivers and the coast intensified in south-east Asia when the rain forest closed up much of the hinterland and that may have inadvertently promoted the island hopping that got people to Australia. Some of these people, it seems, continued the dispersal northwards along the coasts of the western Pacific Ocean. In time they went all the way, past Japan, to Beringia. Both lines of humans, the steppe people and the coastal ones, had descended from a common stock somewhere back in India 25 or more thousand years earlier and now they were in the same area but we do not know whether they met.

This possible meeting of two separate lineages of Ancestors is reminiscent of the meeting between the woodland and steppe people across the mid-latitude belt between France and China that we have already seen. If this view is correct, then it would have been pioneer Ancestors, and not plains people, that came into contact with archaic groups: in Eurasia it is they who would have met the Neanderthals but, with them, they went extinct in many places, leaving few surviving genes. In south-eastern and eastern Asia they may have met the remaining populations of *Homo erectus*. Either way, the Ancestors of the steppe–tundra hardly, if ever, met the archaics, as these had vanished by the time they arrived. Instead they met earlier versions of their own stock. Invariably they seem to have eventually overwhelmed the others—maybe their behavioural strategy, which allowed for more rapid reproductive output, simply turned it all into a numbers game. The irony is that, if this was indeed the case, we may have an example of Ancestors swamping or out-competing others of their own kind, not because they were better or more intelligent but because circumstances had given them a way of life that generated more numbers. Later in history, farmers would swamp-out hunter-gatherers from many parts of the world in a similar fashion.

The steppe–tundra people who were in Beringia by at least 28 thousand years ago, when the last Neanderthals were living in Gibraltar, would have got there by dispersing into areas rich in the kind of fauna that had attracted these people since they first found ways of dealing with them out in the open. Great areas of Beringia that are now the sea bed of the northern Pacific rim would have been steppe–tundra rich in mammoths, reindeer, and wolves. For a long while these people were trapped there as ice sheets and ice deserts hemmed them in on both Asian and North American sides. Then a coastal corridor opened up on the Pacific north-west coast of America as temperatures warmed up and, from an estimated founding population of fewer than 5,000 people, a colonization of North America started between 16 and 15 thousand years ago.[16] It probably followed the coastal route, which opened up a thousand years before an ice-free inland passage appeared between the two main North American ice sheets. From here people would have filtered inland along river systems that cut across the Rockies while others kept within the coastal lowlands between the Pacific and the Rockies, spreading southwards towards South America.

By 14.6 thousand years ago, they had settled at Monte Verde, Chile, where they relied on seaweed and other coastal resources for food.[17] If the dates of entry are correct, approximate though they are, it means that people reached Chile from north-west America in around a thousand years. That puts the rate of spread of this population, according to the crude figures that we have used before for the geographical expansion of humans, at an extraordinary 260 kilometres/generation; that is nearly three times as fast as the spread across the steppe and four times as fast as the expansion from Africa to Australia. It is also the first time in human history that a geographical range expansion went from boreal climates to tropical and equatorial and back to boreal. It is difficult to see people reaching South America so fast if they had had to adapt to the many ecological changes across such latitude bands. It seems likely, instead, that they stayed within a single environment—the coast and its edge habitats inland.[18] In this way they would have been able to maintain a mixed diet and the kind of semi-nomadic existence of their ancient Siberian ancestors, replacing the open steppe with the sea.

In North America, offshoots of this coastal population penetrated inland to find plains similar to those of Siberia and a fauna to go with it. It is probably in response to this newly found resource that the famous Clovis culture emerged. We first pick up this culture that has often been linked with the exploitation of mammoths and mastodonts, between 13.2 and 12.8 thousand years ago;[19] that is, between one and two thousand years after people had settled in Chile. But the North American Plains were exploited soon after the entry and long before Clovis, probably as far back as 14.8 thousand years ago when people were already butchering mammoths on the southern edge of the ice sheets.[20]

Put together, the North American evidence seems to indicate a single and rapid entry into North America via a coastal, ice-free, corridor after a long sojourn in ice-free Beringia. This happened some time between 16 and 15 thousand years ago. Once in, some people kept to the coast, reproduced very rapidly, and spread along the rich coasts using watercraft. Others filtered eastwards across the Rockies and discovered the equally rich Great Plains. As with the first entry into the plains of Central Asia 15 thousand years earlier, or into the Australian Plains 35 thousand years earlier, these people experienced ecological release as they entered a world that had never seen a human before. Here the population grew rapidly and spread right across North America, developing new and highly specialized weapons.

It was the Central American forest that probably acted as a barrier to the further southward dispersal of these people. The Ancestors, who had started their career never far from trees, ironically now found them an impediment. The southern and eastern limits of the vast area of plains, once reaching from France to Beringia and across into North America, were the Atlantic Ocean and the rain forests of Central America. The treeless world, which spanned close to 18 thousand kilometres at its height and had been one of the cradles of humanity, reached its geographical end here.

The world slowly started to come out of the Ice Age after 20 thousand years ago and the first interval of real warming, after 15 thousand years ago, coincided with the entry of people into North America. Further cold came with the brief episode known as the Younger Dryas, when the northern world was plunged back into the Ice Age between

12.9 and 11.6 thousand years ago,[21] but the recovery continued after 11 thousand years ago and the warm world of today had taken shape by 10 thousand years ago. The relentless expansion of the treeless landscapes that, together with the ice sheets, had held Eurasia and North America in their increasingly powerful grip for close to 30 thousand years, now ceased. It was time for the woodlands and forests to hit back. Steppes retreated to Central Eurasia; prairies remained in central North America; the tundra detached from the steppe to occupy a narrow band south of the Arctic; and the ice retreated to the safety of the Arctic Circle.

We will take a look at this new world in the next chapter. We have only briefly entered it here as part of the human colonization of the last major land masses, other than Antarctica, of the planet. The entry was unorthodox, going from north to south when climate was improving. All previous climatic improvements had tended to produce south to north human range expansions. So when people started to spread into the North American treeless landscapes they were paradoxically entering a world that would soon start to shrink. As sea levels started to rise, the coastal people would also see the wide coastal shelves disappearing in front of their eyes. It seems unlikely that humans entered the Americas before or during the Ice Age and we have to assume that, at its height, the Americas were empty of people. So, tropical regions of Africa, Asia, and Australasia aside, where did people survive the height of the last Ice Age and the subsequent Younger Dryas cold pulse?

In the west, the descendants of the Gravettians, with a developed technology that emphasized small, portable, stone armatures, survived in the plains south of the ice sheets. Undoubtedly their sophisticated technology and social systems enabled them to survive this harshest of periods. They managed to do so from Italy in the west, right across the eastern European plains into areas north of the Black Sea.[22] The Central Asian deserts further east probably marked their limits. It seems that, so long as there were grasslands with food, these people could survive the cold and the long winter nights. In Iberia and south-western

France we find a different culture known as the Solutrean, from a site in France. It is thought to be the product of people descended from the Gravettians and its distinctiveness comes from the exquisitely shaped flint arrowheads. The Solutreans may have been among the first people to use the bow and arrow. They had wonderful artists among them who painted the animals of the cold world, especially the horse and the steppe bison, but they also penetrated deep into the south where their numbers swelled beyond recognition. The big modern human demographic explosion in the south of the Iberian Peninsula did not come with the Gravettians, who barely made an impact down here, but with the Solutreans.[23]

Some time around 21 thousand years ago, just as the world was freezing up, a group of Solutreans entered a cave in the very southern limits of Europe, in full view of Africa. They settled in the cave and started a fire in the deep part where they were protected from the wind and the constant blasting of the sand from the dunes outside. Here they were also safe from hungry hyaenas and wolves and from here they made forays into the outside world where they hunted deer and ibex, and collected shellfish from the coast; they caught birds, rabbits, and seals and they scavenged beached dolphins. They took the seeds of the pine trees. The Solutreans were living in Gorham's Cave, Gibraltar, oblivious that other people had lived here for millennia before them. It was now maybe five thousand years since the last Neanderthals had lived here but the new group of people seem to have chosen a similar spot to start their fire: it was the optimal spot where the smoke rose and did not pollute the cave. And they hunted the same animals: what else would a human do if not exploit fully what was available? If it was not for their different build and tools you would not have been able to tell whether the people in the cave were Neanderthals or Ancestors.

One other thing did separate them, but it was cultural, of the kind which also separated different populations of Ancestors. The Solutreans had kept the front teeth of a number of different red deer that they had hunted and one member of the group, a skilled craftsman perhaps, would stay behind in the cave carefully polishing the teeth and drilling perforations into them. These people had carried the Gravettian traditions with them. Here the cave substituted the mammoth bone tent but it was part of a base camp, a place where some people stayed behind and where others came back to exchange intelligence. Another of their group was tasked with a different skill. She was an expert at painting and the wall of a deep part of the cave was turned into a panel in which a red deer, which had contributed to the necklace, was painted. As if wanting to leave an indelible mark of her work of art, she put her hand on the wall and sprayed paint over it. Taking her hand off, its imprint was left for an archaeologist to discover 20 thousand years later.

The Solutreans thrived in Iberia at the height of the cold but so had Neanderthals on many other cold moments long before them—there was nothing too special about that. At the other end of the land mass, in eastern Asia, some hardy people had managed to eke out a living in the narrow pockets of steppe–tundra that survived between the Arctic and Himalayan ice sheets; and there was, of course, Beringia, where people managed to make a living out of reindeer and mammoth. Their chance would come with the thaw. People survived down the Pacific coast of Asia, in Japan where fishing economies became prominent, and in grasslands and woodlands along the Yellow and Yangtze rivers. These populations had connections with the tropical south, unlike the western Eurasians who were cut off from Africa by the Mediterranean Sea and the Sahara Desert.

This was the world of the Ice Age. The people who had survived did so by becoming adaptable to change. In the tropics, Ancestors in Africa, India, and Australasia had no real need to change their ways of life and continued the low-density hunter-gatherer ways that were the patrimony of all humans. That would be one way of being modern that would persist to the present day, albeit in remote outposts as industrialist descendants of the Gravettians, who had lost their own way and all sense of their Pleistocene heritage, insisted on changing the ways of others who had been so successful since millennia before the

Ice Age.[24] The descendants of the steppe–tundra people, including the Gravettians, had found a new and dangerous toy. They had found ways of producing surplus, something almost impossible in warm climates, and with it emerged an unstoppable drive to increase rapidly in numbers. The Ice Age stopped them in their tracks, but only for a while. The Solutreans and others retained the memory and the traditions. As the world warmed up they would put this inherited knowledge to devastating use.

Jared Diamond has eloquently demonstrated how the technological and cultural differences produced by the histories of the peoples of the different continents in the past 13,000 years led to the inequalities of the modern world.[25] The story of chance and historical contingency, which gave some people an edge over others when there were no differences in intellectual capacity between them, did not, however, start 13,000 years ago. It started a long time before that. The entire history of humanity is full of such chance events. The expansion of a population onto the steppe–tundra was another example of an inventive people living on the margins of other, more successful, ones. Need generated invention, and invention generated success when the roulette of life favoured the inventive. The remote Central Asian steppe marked the beginning of a blind experiment that has not yet been completed.

Among the early outcomes of the experiment was the discovery of a new world by these forever opportunists. Because the world they discovered remained even as the climate warmed, the people of the Americas fell into the trap of returning to the old ways. Like the Australians and the Africans these Americans now settled into a hunting and gathering existence. But back in parts of Eurasia, warming removed the steppe–tundra and its animals. The descendants of the Gravettians had to improvise, become inventive once more. For a while they got away with mixed small game and plant economies until, one day, the surplus economy took a radical turn. Some people started to keep grazing animals and shape them while others started to grow plants for food; perhaps some people managed to do both or adopted each other's discoveries. As the woods closed in, trees would be felled to create artificial grasslands for the new grazing animals, just as other humans in a remote past had cleared the rain forest with fire. For now, this was the world of the future. As the Ice Age gripped the world nobody could have seen what was coming.

10

The Pawn Turned Player

IN the previous chapter we saw how people sought refuge from the Ice Age in the south of Europe. They were descendants of the Gravettian people that had spread right across the Eurasian steppe–tundra at a time when increasingly cold and dry conditions were opening up many areas formerly taken up by woodland. The genetic mutations that allowed us to trace the spread of these people from Central Asia also show that an offshoot of these people penetrated southwards across the great mountain ranges of the Caucasus, Zagros, and Hindu Kush into India and western Asia, including the Middle East, some time around 30 thousand years ago.[1] These people who settled south of the mid-latitude belt of mountains had a common heritage with those of the steppe–tundra to the north of these ranges. In the Middle East they may have met the descendants of the early pioneers that had made tentative incursions into Europe but by this time the Neanderthals were long gone.

The Middle East between 30 thousand years ago and the height of the Ice Age around 22 thousand years ago was a junction between peoples from different regions, although the picture provided by the archaeology is far from clear.[2] The people who had been making the Aurignacian in Europe appear to have entered the region too around this time but we cannot be certain who they were or what they looked like; and then there may have been others who had been living there since the early incursions from the expansion that led people to Australia. There was more than one culture in the area at the same time during this period, which seems to support the idea that the people living in the Middle East prior to the Ice Age did indeed come from different sources. The important point is that people from the steppe–tundra got across the mountains and settled in the Middle East, and they are likely to have brought some of the ideas and plains technologies with them.

When the Ice Age hit western Asia, small flint tools which had multiple uses and were much smaller than the blades of the Gravettians became widespread. These microliths were part of a lightweight, portable kit that suited mobile hunter-gatherers across desert, forest steppe, and woodland.[3] The development of such lightweight micro-technologies seems to have taken off during and after the glacial maximum in many regions of Eurasia, although some early attempts appear as far back as 39 thousand years ago.[4] In fact such microliths appear in most Upper Palaeolithic cultures but have often been overlooked because of their small size.[5] They formed part of a general trend, a new fashion perhaps, towards making weapons and tools that could be carried around over long distances and into places where suitable flint sources might not be found. Microliths were part of the trend towards developing risk management strategies that we started to see among the people of the steppe–tundra. Whether they represented independent solutions to similar problems or instead the ideas were transmitted from place to place must remain an open question.

We may not know whether the people of western Asia at the height of the Ice Age adopted ideas from Central Asian immigrants or by independent invention but their behaviour is reminiscent of the people of the steppe–tundra. There were no woolly mammoths or reindeer here but there was a widespread forest–steppe with its own suite of grazers, most notable among them being the gazelle (see Chapter 3). The range of grazers varied across a patchwork of landscapes that ranged from Mediterranean woodland, through forest–steppe to open steppe and desert; there were also gradients of habitat generated by altitude where mountains met lowlands and also islands of habitat within the desert close to oases, as at Azraq in present-day Jordan.[6] Among the animals available to human hunters in this wide range of environments were fallow deer, wild ass, wild goat, aurochs, and wild boar.

The parallels with the hunters of the steppe–tundra are clearly seen at the 23-thousand-year-old site of Ohalo II on the shores of the Sea of Galilee, a submerged site discovered in 1989 following a dramatic fall in sea level. Ohalo II was a village of brush huts, with grass bedding where people slept.[7] The inhabitants of the village were the contemporaries of the Epigravettian people, who in the eastern Mediterranean were the descendants of the Gravettians, and Solutreans of south-western Europe. This was a settlement of hunter-gatherers with structures not made from mammoth bone and hides but from the locally available wood and plants. They were using the raw materials at their disposal just as they were hunting the locally available animals but, local products apart, their behaviour was very similar to that of the Gravettians and their descendants on the other side of the mountains. Chance, once again, would soon bring these people to the brink of a new world; this would be a world inaccessible to the people of the steppe–tundra.

The people living in the steppes of the Middle East were exposed to a range of plants including some of the descendants of the C_4 grasses that, we will recall from Chapter 3, had entered the world millions of years earlier. These were wild cereals and they were harvested by the people of Ohalo who would beat the plants with sticks so that the grain fell in baskets made from plant fibres. Like the Gravettians they were able to make baskets and many other items from plants fibres but, unlike them, they also had access to wild cereals—one day this would become a potent combination.

For now they behaved in much the same way as generations of hunter-gatherers, including the Neanderthals and the Solutreans in far away Gorham's Cave (Chapter 9), had behaved in the rich lands of the Mediterranean: on the shores of the Sea of Galilee they hunted grazing mammals, smaller mammals, birds, and reptiles; they harvested wild plants; they fished; and they beach-combed. But we also find hints of behavioural novelties that we have already observed among the Gravettians: apart from the dwellings and semi-sedentary lifestyle, exotic shells appear inland, suggesting trading networks may have been in operation.

So as we hit the height of the last Ice Age, 22 thousand years ago, we find people on either side of the Mediterranean practising hunter-gatherer lifestyles that were comparable to those of the Mediterranean Neanderthals. But, at the same time, each had its own idiosyncrasies that set them apart from each other and the Neanderthals. In the west people lived and painted in caves while in the east they made huts; in the west they made necklaces from deer teeth and marine shells while in the east they developed exchange networks; crucially, in the east they had something that was absent from the west—wild cereals.

Climate between 22 and 14.7 thousand years ago was erratic and slowly getting warmer as the Ice Age relented. It was after 14.7 thousand years ago, however, that we observe a distinct climatic warming that lasted for just under 2 thousand years. In the northern hemisphere, away from the tropics, this was followed by an abrupt return to Ice Age conditions between 12.9 and 11.6 thousand years ago (Younger Dryas).[8] The next episode of global warming peaked 9 thousand years ago and we have not returned to glacial conditions since.

The first of these periods, between 22 and 12.8 thousand years ago and culminating with the global warming event at 14.7 thousand years ago, saw the establishment of people on all continents except Antarctica. This was still a world of hunter-gatherers although the strategies employed and the species taken varied from region to region and within each region as climate changed. The most dramatic changes took place where climate change was most pronounced, especially away from the tropics. In Europe people spread north from the southern glacial refuges. North-west Europe was colonized by the descendants of the Solutrean people of the south-west.[9] As temperatures soared the ice sheets shrank and the tundra started to recede from the south and expand in the north into areas that had been under the ice. For a while hunters moving north experienced a cornucopia, the last time they would do so, of large grazing mammals. Reindeer and horse were the main species that formed large gatherings at particular sites on the European plains where people gathered and organized communal hunts.[10]

These north-west European hunters also carried with them many of the traditions that had been handed down since Gravettian times, before the Ice Age. For a while they were living in similar contexts to the people of pre-Ice Age Europe. They continued to live in caves where their art flourished but others lived in camps where they made tents from wood and animal hides. The manufacture of statuettes, figurines, and other items of mobile art were also continued. In addition to hunting the large animals, small game such as arctic hare was taken for food and furs and formed part of long-distance trade networks. The similarity in the behaviour of the post-glacial and pre-glacial hunters shows us that the traditions established by the Gravettians had not been lost by the people who lived in the southern refuges.

Although not as well documented as in western Europe, similar changes were happening across Eurasia. In southern China, people hunted elephants, tapirs, deer, and pigs in woodland. Further north, as woodland gave way to open habitats hunters, who wore sewn clothes made from animal hides and furs, relied on the annual slaughter of horse and deer. Further north still, groups of hunters resettled right into the High Arctic zones of Siberia (Chapter 6) and Beringia (Chapter 9) where they relied on the exploitation of the remaining woolly mammoths as well as the ubiquitous reindeer and other animals of the open tundra.

We observe similar lifestyles, with regional and local variations, in newly colonized areas in the Americas. In North America the Clovis culture became established during the period of global warming 13.2 thousand years ago and spread rapidly (Chapter 9). Although these people have been associated with the hunting of the large mammoths, their dependence on these large animals and the extent to which they hunted them remains an open question; small game hunting and plant gathering was regularly practised by Clovis people but art does not seem to have been part of their behavioural repertoire. A similar patterning of hunting and gathering had been practised by the Neanderthals for much of their existence many thousands of years earlier but, in their case, the absence of art has been taken to signify biological inferiority. Such arguments are not put forward when looking at these Ancestors. Quite rightly so: we cannot make inferences about the potential abilities of humans from the presence or absence of art among cultural groups; but such judgements are equally biased when considering Neanderthals as when looking at Ancestors.

To understand the nature of art we clearly must move beyond the biological comparisons, as clearly not all Ancestors painted. While, or even long after, people were painting the wonderful works that we find in such caves as Lascaux in France and Altamira in Spain, others in other parts of the world do not seem to have felt the need to do so. Even those who were painting so beautifully seem to have faded away as global warming transformed the lands of western Europe; the tundra animals vanished and with them their representations on the walls of caves. Art was lost.

Further south from the Clovis territories people had become established within the rainforest of the Amazon shortly after and they painted in caves while becoming reliant on a mixed economy that exploited the strengths of the rainforest. These people fished in the rivers of the Amazon and gathered nuts, roots, and tubers from the hinterland. In the extreme south the inhabitants of the windswept Tierra del Fuego persisted as hunters, to be described one day by a young Charles Darwin. In the Puna grasslands of the Peruvian highlands other people were specialist hunters of the Vicuña while on the rich Pacific what had been a seasonal mixed strategy of exploiting coast and foothills became a full-time fishing economy. These people erected temporary villages on the shoreline and moved between them as the fish shoals moved up and down the coast.

Overall, Ancestors were displaying the adaptability and range of behaviours that had characterized their pre-glacial ancestors and also the Neanderthals. The main difference, and one that was to become increasingly evident as time went by, was that as populations increased in size and information networks became more sophisticated, these people had a corpus of accumulated knowledge that they could draw from. This process of information build-up became less vulnerable to loss as populations grew but at this stage it was still not foolproof: the extinction of the knowledge and skills of the painters of western Europe shows us how precarious it remained.

The coastal exploitation option seems to have been a favourite among people from very different parts of the world. Once the necessary skills had been acquired to effectively exploit the coast a new world of riches became available. Such practices went back to the Neanderthals and proto-Ancestors (Chapter 7) but technology became increasingly sophisticated after the Ice Age and the coast began to support increasingly large groups of people who established sedentary or semi-sedentary lifestyles. We have seen them on the Peruvian coast but similar experiments were being successfully carried out in south-east Asia, where the dense rainforests never made life easy inland, and Japan and the Mediterranean, where sharp open woodland–coast boundaries encouraged a mixed land–sea economy. Where concentrations of fish and other marine resources were dense and predictable the sedentary way of life seems to have become established. In this respect fish and the open sea resembled grazers and steppe–tundra, where we first encountered such lifestyles.

Back in the latitudes of the Middle East global warming produced ecological richness second only to the tropics. The mosaic of Mediterranean woodland, wooded steppe, lakes, and coast over a short distance, along with the seasonal concentrations of gazelles, cereals, and other plants encouraged hunter-gatherers to become sedentary and develop sophisticated technology that included flint sickles for gathering wild plants and large mortars for grinding seeds. The people who settled in this way became known as the Natufians and they have been seen as the precursors of the earliest farmers. In the open steppe other people continued to follow herds and retained a mobile hunter-gatherer existence: the change from nomad to resident was not a clear-cut revolution and it varied in different parts of the world at different times.

Towards the end of this first period of global warming groups of hunter-gatherers, who had established semi-permanent settlements on the foothills of the Zagros Mountains in northern Iraq, became increasingly sedentary as food resources encouraged them to stay put. It is here, where the foothills met the fertile plains of northern Mesopotamia, that farming villages and towns made their first appearance on the global stage 13 thousand years ago. To the west, in the steppes of north-western Syria, the Natufians seem to have found a way of domesticating rye but with the arrival of the cold and dry conditions of the Younger Dryas, 12.8 thousand years ago, this particular project was abandoned and rye returned to its wild state.

The eight hundred years of cold and drought of the Younger Dryas spelt disaster for many of the communities around the Middle East that had settled down to a sedentary way of life in villages. The whole system collapsed and people once more became nomadic hunter-gatherers. Part of the reason was the success of sedentary life itself which had catalysed a massive growth of the population. This sedentary population still relied on wild products, and resources around the villages became depleted. The droughts that came with the Younger Dryas exacerbated the situation and dealt these communities the *coup de grâce*. Sedentary hunting and gathering ways of life, it seems, were only sustainable in the short term.

The Younger Dryas also had its impact on people in other parts of the world. Many northern European sites were abandoned as tundra returned. In others people returned to fully nomadic ways of following the reindeer migrations, on this occasion though with newly acquired bow-and-arrow technology. Changes were not negative everywhere. In North America the aridity that came with the Younger Dryas turned former forests into prairies and opened up new opportunities for hunters of large herds of grazers.

It is after 11.6 thousand years ago that we pick up the final episode of global warming that led to present-day conditions 2 thousand years later. In Europe people once more followed the northward shifting belts of tundra and trees. By now much of the megafauna was localized and on the verge of extinction and others, such as reindeer, were being pushed into Arctic refuges. There was no potential for agriculture or animal domestication here—there were no suitable species for the job. The only possible option was hunting and gathering what was available and that was a rich reservoir of animals and plants: red deer, roe deer, wild boar, and aurochs were among the large mammals; wildfowl, game birds, hares, salmon, and fruit and other plants formed the major part of the European larder. On the coast were the additional resources of the sea including molluscs. With this kind of diverse ecology and the absence of large migratory herds of mammals, people went back to ambush hunting animals widely dispersed within woodland. This was ironically a return to the ways of the Neanderthals, except that they had the accumulated and acquired knowledge that history had provided. Microliths, arrows, and nets were part of this new repertoire.

The people of the Middle East also had a range of wild animals at their disposal with the return of the Mediterranean woodland, mostly gazelles, fallow deer, wild boar, and ibex on the rocks. What were available in abundance were cereals and other plants with the potential for cultivation. With this final episode of global warming came the return to a sedentary life and, with the inherited traditions of the Natufians, ways of harvesting these wild plants. Within a thousand years crops were being cultivated by farmers and towns, like Jericho on the fertile alluvial plain of the Jordan River, sprung up across south-west Asia. As had happened with the Natufians, these early farmers also hunted around the towns while mobile hunter-gatherers continued to live as they had always done on the open steppe.

After 8.5 thousand years ago drought caused the collapse of the farming economy of the Jordan Valley and people once again returned to hunting and gathering. The fertile valleys of the Tigris and Euphrates in present-day Iraq were left to hold the agricultural baton. Economic growth could be sustained here and it soon led to a new form of human enterprise that we have defined as civilization. Except for the abandoned Jordan Valley, towns had sprung up everywhere across the Fertile Crescent by 7 thousand years ago.

Once adopted, and with a favourable climate, there was no stopping the spread of farmers. Eastwards, we find the first farmers on the plain of the Indus Valley by 9.5 thousand years ago. These people appear to have adopted the complete agricultural package from their neighbours in the west: goats, barley, wheat, woodland clearance for growing the crops, houses with mud-brick walls and specialized areas within, grinding stones, flint blades for cutting the cereals, wooden bowls, stone vessels, and baskets but, like in the west, no pottery. As in the west wild plants also continued to be harvested alongside and wild animals were hunted. Farmers and hunters would have interacted and traded with each other in this changing world. Climate once again set limits on this rapid spread of agriculture.

Further into India the seasonal monsoon regime prevented the spread of wheat, barley, and goats: it was simply too hot and too wet in the summer The agricultural package was adopted piecemeal, taking what were suitable elements and adding new ones. India developed a hybrid economy in which the western techniques of farming were applied to local plants, such as millet, and others adopted from the east, most notably rice from China. A similar selective adoption of the agricultural package occurred in southern Europe. Here the first farmers arrived to find relatively people-free landscapes in Greece about the same time as the first farmers started off in the Indus Valley. They probably arrived by sea and brought seeds, sheep, and goats with them. Farmers and hunter-gatherers lived alongside each other for a millennium or more, the farmers keeping to the flood plains and the hunters to the woods and coastline. Eventually the rapid population growth of the farming communities overwhelmed the hunters who either went extinct or became farmers themselves.

It is a story that repeated itself across Europe as farmers and hunters came into contact with each other. In the Mediterranean farmers and hunters appear to have mixed and local people were selective in the elements of the agricultural package that they adopted. Many lived in caves on the coast, probably because the dense forests inland were largely inhospitable, and they continued to harvest the rich resources of the sea. They reached the western end of the Mediterranean by 7.5 thousand years ago, about the time that farmers spreading inland reached many areas of central and western Europe. The hunting and gathering way of life eventually succumbed across Europe by 6 thousand years ago, except in the far north, where it carried on until 2 thousand years ago.

Elsewhere in the world agriculture and farming appear to have been independently invented with the global warming after the Younger Dryas but somewhat later than in the Fertile Crescent. Farmers were harvesting domestic rice in the Yangtze River in China 9.5 thousand years ago and squash and possibly corn in Mexico 10 thousand years ago. To the south vicuña and guanaco were being domesticated in the Andean highlands 7 thousand years ago along with potatoes on the shores of Lake Titicaca.

In other parts of the world hunter-gatherers continued until modern industrialized societies reached them: across much of North America, in Australia, and in sub-Saharan Africa. Global warming had done little to change their way of life in environments simply unsuitable for farming.

The view that I have presented so far in this chapter represents that 'classic' one of the evolution of civilized societies out of an agricultural template. But at the same time I have suggested that the foundations that would one day lead to sedentary urban societies among certain groups of Ancestors are to be found much earlier, among the hunters of the steppe–tundra of 30 thousand years ago. An excessive focus on agriculture and farming has obscured, to my mind, the underlying causes that made people become sedentary and catapulted population growth. The root cause was surplus and the ability to manage surplus. Hunter-gatherers in the tropics might have found themselves in situations where surplus animals could be killed, perhaps during the annual migrations of grazing herds. But they could have done little with the excess as it would have immediately been exposed to the ravages of flies, hyaenas, and the general ensemble of necrophytes that thrive in the tropics. To manage surplus people needed to be able to store the excess and that only became possible when they went onto the Eurasian Plain at a time when climatic conditions permitted the development of storage systems. The Gravettians had been able to store their surpluses in permafrost, but even after such refrigerators disappeared with the Ice Age the knowledge of how to store surpluses survived and could now be adapted to new situations and circumstances.

We have seen so far in this chapter cases of people settling down to a sedentary or semi-sedentary existence without agriculture. The Natufians and the fishermen of coastal Peru are good examples. In one case it was a diversity of resources while in the other it was sheer quantity that encouraged settlement. And there is every reason to expect that, given the right conditions, structured societies with all the paraphernalia of civilization could have arisen in the absence of agriculture and farming. The salmon fishermen of the Pacific north-west coast of America are a case in point.

With the rising sea levels around 8 thousand years ago, the Pacific coastline retreated and narrowed. Once the sea stabilized at a new level people were able to return once again to the coast. Salmon arrived to spawn in the rivers along the shoreline in large numbers at predictable times of the year and a new economy developed around this resource. The fish arrived in a short period so that more fish were taken than could be immediately consumed. A classic surplus scenario arose but people found ways of preserving the extra fish, by filleting them and placing them on racks where the sun and wind would act as drying agents. The staple was supplemented by other fish, seals, land animals, fruit, and nuts from the nearby woods. The two elements that potentially encouraged settlement—quantity and variety—combined to create superb opportunities for the hunter-gatherers of the Pacific north-west.

What emerged was a society in which labour was divided within the community. Surplus and storage allowed craftsmen to take up specialized niches. In time complex societies emerged, capable of establishing hierarchies and defending their resources against their neighbours, of attacking them to secure other resources and trading with others, of exchanging their own products for extraneous ones such as obsidian. This example shows that complex sedentary societies do not need agriculture to develop, although the conditions provided by agriculture are especially well suited for the emergence of such societies.

Göbekli Tepe is a very special archaeological site, located in south-eastern Turkey. Syria is just to the south and some of the earliest domesticated wheat comes from this area.[11] The site is claimed to possess the world's oldest monumental structures, dating back to 11 thousand years ago and predating the advent of agriculture. The site contains stone pillars, some up to 5 metres high, with carvings of animals such as lions and foxes on their sides. Whatever interpretation is given to the site it seems clear that the people who made the structures had created a series of symbolic monuments up on a 780-metre-high hill. And they were hunter-gatherers. This discovery has led some to the conclusion that symbolism (and religion) led to the emergence of agriculture and animal domestication and not the other way round. But Göbekli Tepe was an ecologically rich place: the hunter-gatherers who lived in villages scattered across the landscape had access to large populations of gazelle, aurochs, deer, fruits and nuts, and annual concentrations of migratory birds. It seems that the conditions for the establishment of fairly sedentary communities, a prerequisite for monumental architecture to emerge, were met in the hills of south-eastern Turkey shortly after the end of the Younger Dryas. The ingredients—animal carvings, art on rock surfaces—were not new as they were the legacy of Gravettian and later Ice Age hunter-gatherers. They were simply packaged in a new and spectacular way. Provided the richness of the ecology was not exhausted, or if new ways of maintaining expanding populations were found, there was no reason why these societies could not have developed in complexity, just like the salmon fishermen of the Pacific North-west coast did. That agriculture emerged in this region soon after suggests that such alternatives were indeed found.[12]

The transformation from gathering wild plants to domesticating them was not a sudden revolutionary event but took thousands of years to complete.[13] People living in Ohalo II, on the shores of the Sea of Galilee had been harvesting wild plants since 23 thousand years ago and appear to have been grinding wild cereals into flour.[14] Ohalo II provides us with a direct link between the farmers of the post-Younger Dryas world and the hunters of the Ice Age world and shows how the shift to agriculture was, here at least, a gradual process.

One interesting way of looking at the process of domestication is to see it as a kind of mutual association between humans and the domesticates.[15] Looked at in this way domestication is not a complete break with the past but rather part of a continuum of increasing human intervention from predation to genetic engineering. Domestication shares many things with mutual behaviour in which the partners in the relationship each derive benefits. What distinguishes domestication from other mutual relationships is intentionality on the part of humans. It is this intentionality in the human activity of selecting for particular traits in the animal or plant being domesticated that makes the transformation a relatively fast operation when compared with other mutual partnerships that have taken much longer to evolve.

Evolution by natural selection, as envisaged by Charles Darwin, is the differential survival to reproductive age of individuals of the same species. Those individuals that cope best with their environment and produce most offspring ensure the perpetuation of their genes. It is an unconscious process and a game of numbers. If you can somehow make it to reproductive age, then the more successful offspring you produce the more of the available resources can be taken up by your progeny. In domestication, humans have become a major component of the environment. Domesticated plants and animals benefited from the protection offered by their new environment and were able to out-compete their wild ancestors. Along the way, humans themselves became a part of the domestication process. It is well known that, compared to hunter-gatherers, farmers were generally of lower stature, were in worse nutritional condition, and were more prone to disease.[16] That did not matter. Their numbers kept on increasing regardless, over-whelming the competition in the process.

Why did the Neanderthals not become farmers? This may seem a highly academic question, of little practical consequence, but it is not. It is actually a very relevant question because it helps us understand why agriculture emerged in the places and at the times that it did. The advocates of the superiority of the Ancestors have it easy—for them the answer would be that the Neanderthals simply were not clever enough. If it was not that the Neanderthals were mentally deficient, on the other hand, could climate provide the answer? It provides part of the answer—for the greater part of the period that the Neanderthals lived across Eurasia, after 100 thousand years ago, the climate was much harsher and less stable than today so the conditions favouring agriculture, like those during the Younger Dryas much later, were simply not there. But the Neanderthals had been in the Middle East between 130 and 100 thousand years ago when conditions were milder, so why did they not grow crops and herd animals then? We should not forget that there were proto-Ancestors in the Middle East then too and they did not become farmers either. I think that at the time people were much thinner on the ground and sedentary lifestyles were not in vogue. There was nothing pushing people to change their hunter-gatherer ways. There would be no equivalent warm period for another 100 thousand years. Then there were many more people about, they had learnt to live in villages, and so they depleted wild resources much more rapidly than nomads did. When life got tough they had either to return to the wandering life or, if they chose to stay put, had to find ways of earning a living. Out of the melee of mixed economies emerged some that added sowing seeds and herding wild animals. It allowed communities of people to grow and overpower the hunter-gatherers. Humans auto-domesticated themselves, inadvertently while domesticating plants and animals. The pawn had turned player. For 10 thousand years or so it worked, but the world got smaller as the years went by.

Children of Chance

S OME years ago my wife, Geraldine, and I spent many hours travelling across tracks and unchartered roads across the Iberian Peninsula studying birds and plants. We stopped in many remote places,
sometimes for only a few minutes, and took notes. A few years after
the sites had been visited we would return to repeat our observations.
Before setting off to revisit a location, which was just a number in
our computer, we were hard pressed to remember what the place
looked like. We had our data but, having been to so many places,
the image of each merged into an indistinct mass of trees, rivers, and
cliffs.

What struck us each time was how that image changed and sharpened as soon as we got close to the place that we had been to for
only a few minutes, years before. We put this to the test and amazed
ourselves with our ability to predict what was round the next corner.
Often we got it right down to the position of specific trees, difficult
crossing point, or a geological feature. Clearly all the information had
been stored in our brain and it needed cues hidden in the location
to give our memories a jolt. Unconsciously, we had a mental map of
the geography of each place and we had conducted a crude, but eye-
opening, experiment of ourselves in the process.

The idea that large brains and high mental abilities tend to evolve
among animals that need to be able to move around large territories in
complex environments has been around for some time.[1] These animals
need to create space–time maps so as to be able to locate and return to

sources of food distributed in specific locations within a large area. On the other hand animals eating foods, such as leaves, evenly distributed in an environment do not need to retain as much site-specific information about where food sources are located. Examples of animals that range widely for specific patches of favoured foods and have well-developed mental abilities include chimpanzees, dolphins, whales, hyaenas, elephants, parrots, crows, squids, cuttlefish, and octopi.[2]

An alternative way of looking at the question of the evolution of big brains and mental abilities has looked at the social pressures of life in large and complex groups.[3] Animals living in groups in which relationships between individuals each with its own agenda generates tensions and stresses and need big brains to be able to deal with many possible situations that may arise. For each individual the others in the group become part of a complex and rapidly changing environment. This social brain hypothesis, also known as the Machiavellian intelligence hypothesis, argues that the demands of living in complex social groups best explains the evolution of big and complex brains; it is a view that has tended to overrun the geographical mapping one in recent years.

The life history of an animal may also have an important bearing on the development of a large brain.[4] A slow life history appears to be a prerequisite to becoming brainy. So, two otherwise similar species might differ in mental abilities if one has a slow life history and the other a fast one. Life history acts like a filter that only allows the slow ones a real chance at becoming smart. Geographical mapping and Machiavellian intelligence are catapults only if this hurdle has been surmounted first.

Are geographical mapping, the kind of thing that made me recall where trees and other objects were situated in a place that I had only been to once before, and Machiavellian intelligence really that different? I do not think so. Let me tell you why. Both represent the ultimate way in which an animal can deal with an unpredictable environment. I have suggested in this book that people living in marginal environments needed to be the most inventive. We called them innovators. What I was really saying was that these edge people perceived their environment in a different way from those in the stable core. For them the critical resources that they needed for life, be it food, water, or shelter, were less evenly distributed across the landscape than in the centre so it was harder for them to know where the good bits were. Not only were these good patches unevenly spaced out, they were also unpredictable in where they turned up. So the environment of the innovator was patchier in space and time than it was for the conservative.

Any change that allowed the innovators to pick up cues that made them better at pinning down the elusive and ephemeral resources would have been immediately favoured. The less predictable resources became in space and time, the greater the pressure to find ways of improving flexible systems of detection that could allow for quick reaction. It does not take too much of a stretch of the imagination to see how brainy animals would be favoured in such situations of extreme unpredictability. Should it come as a surprise to find that the brainiest animals, as we have seen, were those that ranged widely for specific patches of favoured foods? The higher mental capacities achieved, the greater the possibilities for flexible behaviour that reduced the risk of going without a critical resource. A clear advantage would have been gained by going in groups—several pairs of eyes are always better than one, not just to find food but also to avoid becoming someone else's food.[5]

But going in groups imposed a new set of pressures that also had to be dealt with. These were the tensions that constantly put the interests of the individual and those of the group on a collision course. For any individual the other members of the group were either resources (collaborators, sexual partners, offspring) or threats (competitors, cheats). The pressures of living in groups were not unlike those that shaped the geographical mapping mind, but they were more intense for the simple reason that the environment now included other members of the same species interacting in myriad ways in a very confined space and over very short time intervals. For animals with slow life histories and sophisticated geographical mapping minds, the shift towards living in groups and adjusting the mindset to cater for even faster changes when least expected would not have been a difficult one.

We have seen in this book how survival is about working the present as best one can with the experience of the past while uncertain of the future. More often than not a radical change in the environment meant that many beautiful models became obsolete overnight but there were times when a design that had evolved for one purpose became useful for a different one altogether. The flexibility of the ape joints helped them in their day-to-day tree-climbing antics but one day it allowed a descendant ape-like creature to make tools, till the soil, and fly to the moon. The ability to walk on two legs in the tree canopy already allowed for the release of a hand to collect fruit from the flimsiest branches. This ability would one day produce Olympic athletes. A general purpose diet that included meat enhanced survival in marginal tropical habitats. It would one day release a proto-human from this tropical prison that had so successfully confined its ancestral lineage for millions of years. Much later in this long history the cultural and technological achievements of the people of the steppe–tundra allowed a few to survive the Ice Age in refugia from which they recovered to colonize much of Eurasia and the Americas. Inadvertently, many of the skills needed to survive in these hostile environments became useful in a new context and agriculture was born.

Other products were of restricted application. The knuckle-walking strategy that allowed proto-chimpanzees to move on the ground between trees in a fragmenting forest took them to a cul-de-sac and surrendered the terrestrial world of savannahs and plains to the humans. Some designs became useful for a long time but eventually succumbed. The spectacular physique of *Homo heidelbergensis*, capable of despatching the large grazers of Middle Pleistocene Eurasia, needed trimming as these animals started to disappear. The Neanderthal was a tweaked version that managed for a while but eventually collapsed along with the remaining mega-animals of this era. It is a lesson to remember. Most designs, perhaps all, given enough time, no matter how perfectly matched to the present they might be, will one day be confronted with the spectre of extinction.

This inadvertent predisposition for future success is beautifully illustrated by the geographical mind of an ancestral ape that roamed a Miocene rain forest; it allowed it to find the trees in fruit when they were in fruit amidst a maze of sterile trunks and branches. That mind became useful in lots of ways, particularly to apes that had been forced to the margins of the forest where good trees were even harder to find. Any change to the brain, which was the engine of that mind, that made it even better at its job was kept and the coded information needed to make this better brain was passed onto future generations of apes. There would have been a limit to all this. That limit would have been set by the cost of producing an increasingly expensive brain. But if the costs could be met by some change that allowed greater intake of energy or by some way of amortizing the cost over the long term, then the expensive brain became a viable proposition.[6] Animal meat, fat, and marrow seem to have provided the currency needed to kick-start brain growth in our ancestors.[7]

By the time that we reach the common ancestor of Neanderthals and us, maybe 600 thousand years ago, we find people that had built on the geographical mapping mind and were polishing a Machiavellian one. This only worked because new pressures and solutions had nudged large brains along.[8] From this common source Neanderthals and proto-Ancestors went separate ways, achieving different, but parallel and comparable, minds. Neanderthals and proto-Ancestors actually reached greater heights when it came to brain size than the Ancestors themselves and they did so by growing the brain at a faster rate while not stopping the period of growth earlier. Their life history was slow, however, perhaps even slower than in the Ancestors,[9] which means that they had all the necessary prerequisites for developing high mental abilities.

There is a clear paradox here: if the Ancestors, our ancestors, are attributed with all the great achievements and revolutions including agriculture that set them apart from all that came before, why did their brains get smaller?[10] One reason may be that they were under greater risk of early death which put pressure on speeding up the life cycle of these people so that they reached maturity and could have offspring sooner; such a scenario has been used to explain the small stature of human pygmy populations.[11] But we have no evidence that the Ancestors were under such pressure so could there have been another cause? One intriguing possibility is that it may have had to do with growing smaller but more streamlined brains that were cheaper and more efficient to run. If so it would have released energy that could have been invested in having babies.

The Neanderthals and their neighbours up to around 30 thousand years ago achieved the largest cerebral hemispheres relative to the cerebellum of any primate, including the Ancestors.[12] As the brain grew increasingly larger during the course of human evolution, towards this peak, the need for it to manage and process data efficiently became greater until a point was reached that may have required the reorganization of its systems of data management, according to one view.[13] The development of the cerebellum in the Ancestors may have been related to having to cope with an increasingly complex cultural and social environment. It would have provided the necessary computational efficiency to work smoothly in the new environment of increasing population density and social and cultural complexity.[14] As with so much of our history this re-organization of the brain that allowed people to interact with each other and their wider environment in increasingly fast and flexible ways was inadvertently predisposing some humans to the world of complex technology that lay ahead.

The reason that we are here and many others are not has to do with numbers. We have seen how hunter-gatherers tended to be physically healthier than their farmer neighbours and yet the latter eventually overwhelmed the former by sheer force of numbers. Sometimes it took longer than others, showing that, where conditions were favourable for a hunting existence or unfavourable for farming, the changeover was not automatic. Jared Diamond has argued that once people went down the agricultural route there was no return to the old ways.[15] This is true within the limits of the unusually stable and warm climatic conditions of the past 10 thousand years but it would have been a different matter had we become re-immersed in another Ice Age. The short cold pulse of the Younger Dryas halted the experimental trend towards the development of crop domestication in the Fertile Crescent, showing us that the agricultural one-way ticket was heavily dependent on having the right climate. The vertiginous post-agricultural expansion of the world population is one example of the kind of reaction that human populations had when a change in either biology or culture released them from the ecological constraints that were checking their numbers. But the flood gates had opened many times before, though admittedly not producing outcomes of this magnitude.

The major demographic and geographical surges in our history have had to do either with changes in the biology (and later on also culture) of species or with the breakdown of ecological barriers. We first met such a change during the global warming period in the Eocene that allowed early primates to spread across the forests right up to what is now the Arctic. In this case it was climate change that created the opportunity for forest expansion and released the small primates from their equatorial homes. At other times it required a combination of biology and environmental change to get things going. This was the case with the Miocene apes that spread across the subtropical forests of Eurasia. As during the Eocene, a period of climate warming allowed the spread of the warm, seasonal, forests across Eurasia. But the apes trapped in Africa could not get to them for two reasons: on the one hand the sea separated Africa from Eurasia so they could not get across and on the other they did not have the right biology, in this case tough teeth for eating hard nuts, to subsist in these seasonal forests. Once both problems had been overcome the apes rapidly spread across Eurasia and diversified into many distinct forms.

There have been a number of occasions during the long course of our evolution when a change in the biology of a species has allowed it to expand geographically over a wide area.[16] As the proto-humans w - Com

Sma Dryer olive

I rel new

widened their habitat tolerance, with corresponding changes to their anatomy, they spread across tropical Africa from a core area in present-day Ethiopia, then south into South Africa. Quite possibly this change allowed them to move north and into parts of Eurasia also. The rapid expansion of *Homo erectus*, around 1.8 million years ago, across the savannahs of Africa and southern Eurasia is also attributable to a biological change that marked the start of our genus, *Homo*.

Other such rapid expansions have to do with changes in technology and culture rather than hardware although it is difficult to tease out when technology drives the change and when instead it is a change in the environment. The rapid expansion of the proto-Ancestors across North Africa and into Arabia seems to be linked with a period of aridity that opened up huge areas of former woodland into savannah and semi-arid steppe. The people who had been living in a restricted area of north-east Africa in such environments suddenly found an opportunity as their habitat expanded. They carried an early version of projectile technology, the Aterian, that may have synergized with the climate-driven environmental change to speed up the process.

This kind of ecological release, allowing for rapid gains of territory, seems to have marked the major human expansions into unpopulated regions of the world. From Arabia Ancestors moved into India first and then across south-east Asia following belts of savannah that opened up as climate changed. Once inside Australia the spread of people across a continent empty of humans proceeded at lightning speed. The great expansion of Ancestors across the steppe–tundra of Eurasia again was by people who had adapted to this habitat in a localized area. As with the Aterians, once the habitat spread people followed as they tracked their favourite food resources and habitats. The entry into North America, once the Bering barrier was lifted, was among the most rapid in prehistory. New boat-building technology may account for the incredible speed with which the southern tip of South America was reached while others found, as the first Australians had done, a vast empty land of prairies and savannahs with a bounty of grazing animals.

These rapid gains of territory were soon brought to a halt as populations settled down at densities that could be supported by the new environments. Although there are claims of over-exploitation and even mass overkill of animals as Ancestor populations spread,[17] there is little to support the contention except in the confined spaces of islands. But even if such overkill did take place in particular locations, there is no evidence of massive population explosion of the kind that we see with the start of food production. It is with the farmers that we see the big change in population density and social structure. Happening as it did, once climate had stabilized, the demographic and geographical expansion of farmers had much more to do with new technology than with a change in the environment. It marked the start of the illusion of progress towards a world of unsustainable growth, a dream that has turned into a nightmare as we procrastinate today while the current state and the future of our planet hang in the balance as a result of our voracity. How could we have reached such an unhealthy state of affairs? The answer lies in the way in which we got to the present, not as evolutionary superstars but as pests that invaded every nook and cranny that became available.

Taming the future is the essence of the human story. Recall that the successful populations that ultimately led to us were always those living on the edge of others who monopolized the good-quality territory. We were born from the poor and feeble that had to spend every ounce of energy searching for the scraps that kept them alive. This may seem a little undignified for those of us who see ourselves at the pinnacle of evolution but that is the sobering reality of our story. Every step of the way in the unpredictable story that led to us was marked by populations of innovators living on the periphery. Many fell by the wayside but one population made it to tell the tale.

Who were these edge innovators that contributed to our lineage? There were many: the Miocene apes that started to eat leaves and nuts in seasonal forests because they did not have access to the prime rain-forest and its wealth of fruit; Toumaï or one of its cousins who started to experiment on the edge of these forests; Ramidus, Lake Man, and their kind who ventured onto the periphery of the multilayered forest canopy; *H. erectus* and his descendants, including *H. heidelbergensis*, who lived where woodland met grassland and took meat-eating to a new level; the Nubians and the Aterians who learnt how to deal with the inhospitable desert; the Niah people who managed to survive on the edge of the daunting rainforest; the survivors of the Toba eruption; and the Central Asians who adapted to the steppe and made the most of it when the steppe–tundra overwhelmed Eurasia. If we went back in time, without knowledge of how climate would change the world, we would not give any of these guys a fighting chance. Surely they would all have been weeded out by natural selection. But they were not, and we are here because of their resilience and their luck.

By living on the edge the innovators had to keep finding ways of reducing the risk of going without food, water, shelter, or mates. It is my contention that life in the marginal territories sifted out the inventive individuals from the rest. These super-survivors could deal with the risk of an unpredictable supply of food or water better than any others of their kind so that when climate changed and made matters worse all round, it was they and their offspring who fared best. The earliest form of risk management seems to have involved living on the edge of two or more habitats or in a patchwork of habitats. Once away from the forest comfort zone these innovators fared best by sticking to places where several kinds of habitats were close by and this allowed them to exploit a wider variety of foods than if they lived in a single habitat. It is a strategy that seems to have been long lived: we find Toumaï already living on a lake margin close to the edge of woodland; we find the strategy again, later, among the Mediterranean Neanderthals who lived between cliff, savannah, lake, and coast; the Neanderthals and proto-Ancestors of Skhul, Qafzeh, and Tabun who lived in habitat mosaics and savannahs; or the Ancestors in Niah who lived on the edge of rain-forest, river, and savannah; and we even find it among many surviving groups of hunter-gatherers that reached our time in places like Australia, the Kalahari, and the Americas: these were the smartest humans of all.

The parallel tendency to diversify the diet, first increasing the range of plant foods and then adding a range of animal foods, was another way of reducing risk by not putting all one's eggs in one basket. Archaeologists have tried to find moments in our evolutionary history when the range of foods eaten by people started to widen—they have called this the broad spectrum revolution.[18] This is one more revolution that, in my view, must be consigned to the waste-paper basket along with all the other purported revolutions of the human story.[19] There was no point in human history when people across the globe decided to eat a wider range of foods. A mixed diet was always part of our biology. It was a good way of minimizing the chances of going without food and it characterized the way that our bodies were shaped in order to be able to process a wide range of items. What did change was what was eaten and that, like today, varied from place to place according to what was available. If we compare the Ancestors of Ice Age Mediterranean Iberia, the Middle East, the earlier Mediterranean Neanderthals, the Clovis people of North America, the early hunter-gatherers of the Amazon, and the hunter-fishermen of coastal Peru, we see parallel mixed economies but each with its own idiosyncrasy that had to do with location and tradition. A general-purpose habitat tolerance and diet were the early ways of managing risk. They were so successful that they stayed with us until today. A mixed economy was one essential component in the switch to agriculture and farming in many communities across the world. Mixed economies were occasionally substituted by specialized ones in which high returns from a single product made the switch worthwhile. But these quick return tactics tended to be high risk and short-lived as the products became exhausted. In the story of human evolution these strategies were not the rule but when they did become successful they changed the world for ever. The starting point was not the Fertile Crescent of 10 thousand years ago but the Russian Plain and its 30-thousand-year-old Gravettian culture.

The conquest of the Eurasian steppe–tundra by Ancestors around 30 thousand years ago marked a dramatic shift in the fortunes of a population that would swamp the whole of Eurasia and the Americas.

Only tropical Africa, parts of southern Asia, and Australasia avoided the wave of new colonizers. These people, typified by the makers of the Gravettian culture in Europe, developed a way of life that brought together many cultural, technological, and social skills already present in other humans. Their talent was in bringing all these various elements together as a single package. It was a package that defined the people of the northern hemisphere and it was from here that agricultural communities would eventually emerge. The change of circumstances for these people had nothing to do with a sudden, quasi-miraculous, biological transformation. It was simply the consequence of having to manage risk much more efficiently than ever before.

We have seen throughout this book how particular attributes, evolved for particular functions, have become useful in other ways in time. We have spent some time looking at how our brain developed from a geographical mapping organ to one capable of much more. There was a by-product: it was awareness of ourselves, not something unique to us but certainly unique to the group of beings that seem to have developed similar brains under similar pressures. Octopi, cuttle-fish, and their kind appear to have a kind of primary consciousness,[20] and self-awareness comparable to that in humans now seems confirmed in elephants,[21] bottlenose dolphins,[22] and apes.[23] Awareness of self would seem to be a natural consequence of awareness of objects in space and time including other members of one's own species. Having got it, whether because it carried some unknown advantage or more likely a side-effect of developing a large and complex brain, self-awareness was added to our complex and intricate systems of information transfer and communication. It produced an animal capable of situating itself in space and time, an animal that became aware of the consequences of its own behaviour and mortality.

That very self-awareness gave humans their capacity for rational thought, to be conscious of the consequences of their actions, and the ability to remedy those that were detrimental in some way. But, in the same breath, it allowed conscious Machiavellian behaviours and manipulation. Gossip, essentially information and disinformation about others in the group, became central in our daily lives. Signalling that served to upgrade and maintain our status within the group became a survival tactic in its own right. When it came to our neighbours, trinkets and artefacts became trading currency and ways of demonstrating our own superiority. When I turn on the television and see the importance that we have given to reality and gossip shows, and to political spin, and I see how irrationally we conduct ourselves supporting football teams with which we have no real connection, I see a direct measure of how far we have lost the plot.

We have also seen throughout this book how we owe our existence to chance. From asteroid impacts to volcanic eruptions to simply being in the right place at the right time, we are here because of luck. It is easy to fall into the circular reasoning that we are here and therefore we are the product of successful genes. We should not delude ourselves. Our genes are successful, like those of all other species that exist today, only to the extent that they have made it to this point. But as we have seen we are here because of a combination of successful and lucky genes, those that by chance coincided with favourable conditions or were able to keep up with the pace of a changing world. There were many highly successful lineages that went extinct because their luck ran out—the Neanderthals and other populations of proto-Ancestors among them.

More lineages of humans disappeared on the way than those that made it to today. But I am not going to fall into the trap of gloomy predictions about the future of our own species. Our population has reached such large numbers that it is unlikely that any radical change in climate will now ever wipe us out completely. It would take a rapid catastrophe of global dimensions to achieve that. Without doubt many will suffer and perish from floods and famine but those of us in the comfort zone will pretend to care and do nothing about it. This is the sadness of our story—if one thing makes us unique it is our awareness of our actions and our ability to change things if we choose to, but more often than not we do not. Having spent our evolutionary lives trying to cope with change and finding ways of handling and cheating the unpredictable future, now that we have it in our hands to change the future we procrastinate or choose not to. The reason lies in that internal tension in all of us that wrestles between self and our neighbour, between individual gain and the higher gains to be had from working in a team.

Ten thousand years, the time that defines the post-agricultural changes that have brought us to the present, is miniscule when compared to the long story of our evolution. It is a mere 0.2% of the time since our ancestors split from the chimpanzee lineage; 0.6% of the time since the appearance of *H. erectus*, the first member of our genus (*Homo*); 1.67% of the time since Neanderthals split from our lineage; 2.5% of the entire time that the Neanderthals lived on this planet; and around 5% of the time that we ourselves have been around in some shape or form that we might feel confident to call *Homo sapiens*. It is 5 thousand years less than the time that it took the Ancestors to penetrate Eurasia. Ten thousand years is almost imperceptible in the record of our species. It is in this time that we have swayed off at a tangent and lost contact with our biological heritage.

When we pause to consider what a tiny fraction of our evolutionary history has been taken up by our post-farming existence, it becomes blindingly obvious that our biological make-up was formed almost entirely before farming. We continued to evolve after this, of course, even though it probably took the form of a kind of self-domestication.[24] There has been an increasing realization in recent years that our technological and cultural achievements of the past 10 thousand years have drawn us out of the orbit to which we had become adapted—a mismatch has arisen between our biology that took millions of years to evolve and our current lifestyles that took only a few thousand years to develop.[25] Throughout human evolution populations that could not handle rapid rate of change in their environment caused by some perturbation went extinct. The Neanderthals are a prime example. Now it is us, through culture and technology, who have introduced a perturbation at a pace that our bodies are finding hard to match.

We should not forget that we are the product of marginal people who had to do a lot of improvising to get by. Culture and technology offered us the great chance to react to climate and environmental change much faster than our genes could. We took off, modifying and changing the environment and our foods, becoming increasingly independent of that environment and producing more and more offspring. For a while it worked—the world was so big and we were so few that we fell under the spell of our own achievements. It all seemed sustainable—there was no end to available resources—and we kept on going. But at the time scales we are interested in, 10 thousand years is a mere drop in the ocean. As the population of the planet has grown we have become increasingly aware that this particular project is only sustainable at these short time scales and that one day it will all tumble down. We have seen spectacular collapses of apparently unassailable civilizations in recorded history but nothing will compare with what lies ahead.

And when it all comes tumbling down who will survive? There is enough in our story to suggest that it will not be those of us in the comfort zone, the auto-domesticated slaves of electricity, motor cars, and cyberspace, who would not last more than a few days without supporting technology. The tradition that produced the bureaucrat, the priest, and the king generated communities of specialists, which was fine as long as conditions were favourable. But when things get bad these societies of experts will become strained to their limits. The children of chance, those poor people who today must scrap for morsels each day without knowing when and where the next meal will come from, will once again be the most capable at survival. The innovators will once again win when the rapid and powerful perturbation that will be economic and social collapse, generated by the conservatives themselves, will ironically mark their own downfall. And evolution will take another step in some as yet unknown direction.

Endnotes

Prologue—When Climate Changed the Course of History

1. In this book I will refer to all members of the genus *Homo* as humans since all shared a common ancestor with us. That common ancestor was Upright Man, *Homo erectus*. The precise taxonomic category of *H. erectus* is open to debate as it constitutes a direct ancestral-descendant lineage with us. I will refer to the lineage that led exclusively to us as *Homo sapiens*, the Ancestors, as I have explained in the Preface. Because there is some debate on the taxonomic status of earlier, incipient, versions of *H. sapiens*, those from 200 to ∼ 130 thousand years ago, I will refer to as proto-*H. sapiens*, or proto-Ancestors. Those that came after will simply be referred to as *H. sapiens*—the Ancestors.

2. A. Weil, 'Living Large in the Cretaceous', *Nature* 433(2005): 116–17; Y. Hu et al., 'Large Mesozoic Mammals Fed on Young Dinosaurs', *Nature* 433(2005): 149–52; Q. Ji et al., 'A Swimming Mammaliaform from the Middle Jurassic and Ecomorphological Diversification of Early Mammals', *Science* 311(2006): 1123–7.

3. The earliest period of the Tertiary Epoch, just after the asteroid impact.

4. K. C. Beard, 'The Oldest North American Primate and Mammalian Biogeography during the Paleocene–Eocene Thermal Maximum', *Proc. Natl. Acad. Sci. USA* 105(2008): 3815–18.

5. J. Zachos et al., 'Trends, Rhythms, and Aberrations in Global Climate 65 Ma to Present', *Science* 292(2001): 686–93.

6. C. Janis, 'Tertiary Mammal Evolution in the Context of Changing Climates, Vegetation, and Tectonic Events', *Ann. Rev. Ecol. Syst.* 24(1993): 467–500; J. S. Carrión, *Ecología Vegetal* (Murcia: DM, 2003).

7. D. R. Begun, 'Planet of the Apes', *Sci. Amer.* 289(2003): 64–73.

8. D. R. Begun, '*Sivapithecus* Is East and *Dryopithecus* Is West, and Never the Twain Shall Meet', *Anthropol. Sci.* 113(2005): 53–64.

9. *Chororapithecus abyssinicus*. G. Suwa et al., 'A New Species of Great Ape from the Late Miocene Epoch in Ethiopia', *Nature* 448(2007): 921–4.

10. Begun, 'Planet of the Apes'.

11. Begun, '*Sivapithecus* Is East'.

12. Suwa et al., 'A New Species of Great Ape'.

13. *Nakalipithecus nakayamai.* Y. Kunimatsu et al., 'A New Late Miocene Great Ape from Kenya and Its Implications for the Origins of African Great Apes and Humans', *Proc. Natl. Acad. Sci. USA* 104(2007): 19220–5.

14. *Ouranopithecus macedoniensis.* Begun, '*Sivapithecus* Is East'.

15. J. F. Burton, *Birds and Climate Change* (London: Christopher Helm, 1995).

16. J. M. Bowler et al., 'New Ages for Human Occupation and Climatic Change at Lake Mungo, Australia', *Nature* 421(2003): 837–40.

17. C. A. Brochu and L. D. Densmore, 'Crocodile Phylogenetics: A Summary of Current Progress', in G. C. Grigg et al. (eds), *Crocodilian Biology and Evolution* (Chipping Norton, NSW: Surrey Beatty and Sons, 2000), 3–8.

18. L. A. Sawchuk, 'Rainfall, Patio Living, and Crisis Mortality in a Small-Scale Society: The Benefits of a Tradition of Scarcity?', *Curr. Anthropol.* 37(1996): 863–7.

19. C. Finlayson, *Neanderthals and Modern Humans: An Ecological and Evolutionary Perspective* (Cambridge: Cambridge University Press, 2004).

Chapter 1—The Road to Extinction Is Paved with Good Intentions

1. Human hearing differs from that of chimpanzees in having a relatively high sensitivity, from 2 up to 4 kHz. This region contains relevant acoustic information in spoken language. The skeletal anatomy of the Sima de los Huesos people shows that they had a human-like pattern of sound power transmission, through the outer and middle ear, at frequencies up to 5 kHz. These results suggest that they already had auditory capacities similar to those of living humans. I. Martinez et al., 'Auditory Capacities in Middle Pleistocene Humans from the Sierra de Atapuerca in Spain', *Proc. Natl. Acad. Sciences USA* 101(2004): 9976–81.

    Our closest extinct relatives, the Neanderthals, shared with modern humans two evolutionary changes in FOXP2, a gene that has been implicated in the development of speech and language. These genetic changes were present in the common ancestor of modern human and Neanderthal populations. J. Krause et al., 'The Derived FOXP2 Variant of Modern Humans Was Shared with Neandertals', *Curr. Biol.* 17(2007): 1908–12.

2. To simplify matters I will refer to proto-humans as those fossils that show features that implicate them in the human story but cannot be considered to have the full suite of characters necessary that we would consider them to be fully human. Not all will necessarily be our Ancestors. For our purposes these will include the genera *Orrorin*, *Sahelanthropus*, *Ardipithecus*, *Australopithecus*, *Paranthropus*, and *Kenyanthropus*. They are usually referred to as hominins in the scientific literature. I will also refer to all members of the genus *Homo* from *Homo erectus* onwards as humans but consider earlier forms (*rudolfensis*, *habilis*, *georgicus*) as proto-humans. In this interpretation, I recognize the argu-

ments for the alternative designation of *Australopithecus habilis* not *Homo habilis* (B. Wood and M. Collard, 'The Human Genus', *Science* 284(1999): 65–71), but retain the latter as it appears more frequently in the scientific literature.

3. Estimates of the time of separation of the orang-utan, gorilla, and chimpanzee lineages from that of the human vary substantially. Molecular clocks, which compare genetic distance between living species, convert the estimates into time, assuming mutations are neutral and arise at a constant rate. Population size also affects the estimates as do the selected calibration points of the clock, which are usually based on reliable age estimates for known fossils. F. J. Ayala, 'Molecular Clock Mirages', *Bioessays* 21(1999): 71–5; J. H. Schwartz and B. Maresca, 'Do Molecular Clocks Run at All? A Critique of Molecular Systematics', *Biol. Theory* 1(2006): 357–71.

    Estimates for the orang-utan divergence vary between 18 and 11 million years ago (mya); for the gorilla divergence between 8.4 and 5 mya; and the chimpanzee divergence between 7 and 4 mya. R. L. Stauffer, 'Human and Ape Molecular Clocks and Constraints on Paleontological Hypotheses', *J. Hered.* 92(2001): 469–74; F-C. Chen and W-H. Li, 'Genomic Divergences between Humans and Other Hominoids and the Effective Population Size of the Common Ancestor of Humans and Chimpanzees', *Am. J. Hum. Genet.* 68(2001): 444–56; Z. Yang, 'Likelihood and Bayes Estimation of Ancestral Population Sizes in Hominoids Using Data From Multiple Loci', *Genetics* 162(2002): 1811–23; G. V. Glazko and M. Nei, 'Estimation of Divergence Times for Major Lineages of Primate Species', *Mol. Biol. Evol.* 20(2003): 424–34; D. E. Wildman et al., 'Implications of Natural Selection in Shaping 99.4% Nonsynonymous DNA Identity between Humans and Chimpanzees: Enlarging Genus *Homo*', *Proc. Natl. Acad. Sciences USA* 100(2004): 7181–8; S. Kumar et al., 'Placing Confidence Limits on the Molecular Age of the Human–Chimpanzee Divergence', *Proc. Natl. Acad. Sciences USA* 102(2005): 18842–7; N. Patterson et al., 'Genetic Evidence for Complex Speciation of Humans and Chimpanzees', *Nature* 441(2006): 1103–8; A. Holboth et al., 'Genomic Relationships and Speciation Times of Human, Chimpanzee, and Gorilla Inferred from a Coalescent Hidden Markov Model', *PLoS Genet.* 3(2007): 294–304; I. Ebersberger et al., 'Mapping Human Genetic Ancestry', *Mol. Biol. Evol.* 24(2007): 2266–76. Bipedalism is first observed in *Orrorin tugenensis* 6 mya and persisted for 4 mya until modifications in the hip appeared in early *Homo*. B. G. Richmond and W L Jungers, '*Orrorin tugenensis* Femoral Morphology and the Evolution of Hominin Bipedalism', *Science* 319(2008): 1662–5.

4. Estimates are of effective population size, which approximates to the size of the breeding population. J. D. Wall, 'Estimating Ancestral Population Sizes and Divergence Times', *Genetics* 163(2003): 395–404.

5. The fossils have been assigned to a new species, *Sahelanthropus tchadensis*. M. Brunet et al., 'A New Hominid from the Upper Miocene of Chad, Central Africa', *Nature* 418(2002): 145–51.

6. M. Pickford and B. Senut, 'The Geological and Faunal Context of Late Miocene Hominid Remains from Lukeino, Kenya', *C. R. Acad. Sci. Paris, Earth Plan. Sci.* 332(2001): 145–52.

7. Pickford and Senut specifically remove *Australopithecus afarensis*, which includes the famous Lucy, from the human ancestry and also propose that *Ardipithecus* is the ancestor of the chimpanzees.

8. The fossils were initially ascribed to a subspecies, *Ardipithecus ramidus kadabba*, and subsequently elevated to full species, *Ardipithecus kadabba*, in 2004. Y. Haile-Selassie, 'Late Miocene Hominids from the Middle Awash, Ethiopia', *Nature* 412(2001): 178–81; G. WoldeGabriel et al., 'Geology and Palaeontology of the Late Miocene Middle Awash Valley, Afar Rift, Ethiopia', *Nature* 412(2001): 175–8; Y. Haile-Selassie et al., 'Late Miocene Teeth from Middle Awash, Ethiopia, and Early Hominid Dental Evolution', *Science* 303(2004): 1503–5.

   Ardipithecus is a combination of *Ardi*, which means ground or floor in the Afar language of the people of the Awash Region of Ethiopia, and the Latin *pithecus* meaning ape. *Kadabba* in the Afar language means basal family ancestor. *Ardipithecus kadabba* therefore means 'ground ape ancestor at the base of the [human] family'. T. D. White et al., '*Australopithecus ramidus*, a New Species of Early Hominid from Aramis, Ethiopia', *Nature* 375(1995): 88; Haile-Selassie, 'Late Miocene Hominids'.

9. P. Vignaud et al., 'Geology and Palaeontology of the Upper Miocene Toros-Menalla Hominid Locality, Chad', *Nature* 418(2002): 152–5.

10. *Australopithecus bahrelghazali* was named after the Bahr el Ghazal Valley in Chad, where it was found. M. Brunet et al., 'The First Australopithecine 2,500 kilometres West of the Rift Valley (Chad)', *Nature* 378(1995): 273–5; M. Brunet et al., '*Australopithecus bahrelghazali*, une nouvelle espece d'Hominide ancien de la region de Koro Toro (Tchad)', *C. R. Acad. Sci. Paris, Earth Plan. Sci.* 322(1996): 907–13.

11. The skull was discovered in 1924 in Taung, South Africa, and published as *Australopithecus africanus* by Raymond Dart in 1925: R. A. Dart, '*Australopithecus africanus*: The Man-Ape of South Africa', *Nature* 115(1925): 195–9.

12. *Australopithecus afarensis*. D. C. Johanson and M. Taieb, 'Plio-Pleistocene Hominid Discoveries in Hadar, Ethiopia', *Nature* 260(1976): 293–7; D. C. Johanson et al., 'A New Species of the Genus *Australopithecus* (Primates: Hominidae) from the Pliocene of Eastern Africa', *Kirtlandia* 28(1978): 1–14.

13. The exact number of species of small-brained proto-humans varies according to different authors. All lived in the area from South Africa to Ethiopia and west to Chad. The species which have been recognized are: (1) *Ardipithecus ramidus* (Ethiopia, 4.51–4.32 mya); (2) *Australopithecus anamensis* (Ethiopia and Kenya, 4.2–3.9 mya); (3) *A. afarensis* (Ethiopia, Kenya, Tanzania, 3.9–3.0 mya); (4) *A. bahrelghazali* (Chad, 3.5–3.0 mya); (5) *Kenyanthropus platyops* (Kenya, 3.5–3.2 mya); (6) *A. africanus* (South Africa, 3.3–2.3 mya); (7) *Paranthropus*

aethiopicus (Ethiopia, Kenya, 2.8–2.3 mya); (8) *Australopithecus garhi* (Ethiopia, 2.5 mya); (9) *Paranthropus boisei* (Malawi, Tanzania, Kenya, Ethiopia, 2.5–1.4 mya); (10) *Paranthropus robustus* (South Africa, 2.0–1.5 mya); (11) *Homo habilis* (Ethiopia, Kenya, Tanzania, South Africa, 2.33–1.44 mya).

To this series we should add a largely complete, and as yet unnamed, skeleton of an *Australopithecus* from Sterkfontein, South Africa (recovered between 1994 and 1998), dating to 3.33–3.0 mya. It is the oldest *Australopithecus* discovered from South Africa and is not considered to belong to *A. africanus*. The first bones that were found belonged to the foot and, because of the small size of the individual, it has become known as Little Foot. T. C. Partridge et al., 'The New Hominid Skeleton from Sterkfontein, South Africa: Age and Preliminary Assessment', *J. Quat. Sci.* 14(1999): 293–8. The authors have subsequently proposed a date of approximately 4 mya for the fossil but this is controversial. T. C. Partridge et al., 'Lower Pliocene Hominid Remains from Sterkfontein', *Science* 300(2003): 607–12.

14. *Ardipithecus ramidus*, intitially described as *Australopithecus ramidus*. Ramid means root in the Afar language of the region. See n. 8 for etymology of *Ardipithecus*. T. D. White et al., '*Australopithecus ramidus*, a New Species of Early Hominid from Aramis, Ethiopia', *Nature* 371(1994): 306–12; White et al., '*Australopithecus ramidus*, a New Species of Early Hominid from Aramis, Ethiopia, Corrigendum', 88.

15. S. K. S. Thorpe et al., 'Origin of Human Bipedalism as an Adaptation for Locomotion on Flexible Branches', *Science* 316(2007): 1328–31.

16. *Australopithecus anamensis*. Anam means lake in the Turkana language so the full name means southern ape of the lake. M. G. Leakey et al., 'New Four-Million-Year-Old Hominid Species from Kanapoi and Allia Bay, Kenya', *Nature* 376(1995): 565–71.

17. T. D. White et al., 'Asa Issie, Aramis and the Origin of *Australopithecus*', *Nature* 440(2006): 883–9.

18. Johanson and Taieb, 'Plio-Pleistocene Hominid'; Johanson et al., 'A New Species'.

19. The 3 million-year-old skeleton of a three-year-old *A. afarensis* child, published in 2006, suggests that it probably did not walk exclusively on two feet. Z. Alemseged et al., 'A Juvenile Early Hominin Skeleton from Dikika, Ethiopia', *Nature* 443(2006): 296–301.

20. J. G. Wynn et al., 'Geological and Palaeontological Context of a Pliocene Juvenile Hominin at Dikika, Ethiopia', *Nature* 443(2006): 332–6.

21. See Pickford and Senut, 'The Geological and Faunal Context'.

22. The unearthing of a new fossil proto-human is guaranteed to make the news but sometimes other discoveries and new analyses are more sensational, even though they may not receive so much publicity. One such exciting result was reported in 2007 and it involved the examination of a recently discovered jaw of one of Lucy's people. The results were staggering: Lucy and her people could

not have been our ancestors. Their jaws had features in common with other proto-humans that came later but not with humans, nor chimpanzees. In fact, closest mandibles were those of gorillas but the similarity was attributed to independent evolution of these features in the two and not to some obscure evolutionary relationship. When the mandible of *Ardipithecus ramidus* was looked at, it was found to resemble chimpanzees and humans but not Lucy. What this seems to indicate is that *A. ramidus* at 4.4 million years ago could have been on the way towards humans or chimpanzees and maybe even Lucy but shortly afterwards Lucy's branch went one way and the future human branch another. The chimpanzee line may already have parted or *A. ramidus* was indeed very close to the human–chimpanzee split. Y. Rak et al., 'Gorilla-Like Anatomy on *Australopithecus afarensis* Mandibles Suggests *Au. Afarensis* Link to Robust Australopiths', *Proc. Natl. Acad. Sciences USA* 104(2007): 6568–72.

23. All proto-humans of the genera *Australopithecus* and *Paranthropus*. See also n. 13.

24. *Kenyanthropus platyops*: flat-faced man from Kenya. M. G. Leakey et al., 'New Hominin Genus from Eastern Africa Shows Diverse Middle Pliocene Lineages', *Nature* 410(2001): 433–40.

25. *Homo rudolfensis*, discovered in 1972 and named in 1986. V. P. Alexeev, *The Origin of the Human Race* (Moscow: Progress, 1986).

26. The ancestors of the chimpanzee and the bonobo.

27. *Paranthropus boisei* in East Africa and *P. robustus* in South Africa.

28. The oldest known associations between stone tools and broken animal bones are from Gona in Ethiopia and date to 2.6 mya. The makers of the tools are unknown but *A. garhi* is suspected. In any case tool making predates the first appearance of *Homo*. S. Semaw et al., '2.6-Million-Year-Old Stone Tools and Associated Bones from OGS-6 and OGS-7, Gona, Afar, Ethiopia', *J. Hum. Evol.* 45(2003): 169–77; M. Domínguez-Rodrigo et al., 'Cutmarked Bones from Pliocene Archaeological Sites at Gona, Afar, Ethiopia: Implications for the Function of the World's Oldest Stone Tools', *J. Hum. Evol.* 48(2005): 109–21.

29. *Paranthropus boisei*, *P. robustus*, *H. habilis*, and *H. rudolfensis* are described from the early part of the Pleistocene up to around 1.4 mya. The latter two species are surrounded in ambiguity, because of the fragmentary nature of the finds, and may be a single species or an array of forms. They have been traditionally placed in the genus *Homo* but Wood and Collard, 'The Human Genus', place them in *Australopithecus*. See also I. Tattersall and J. Schwartz, *Extinct Humans* (Boulder, CO: Westview Press, 2000).

30. Alexeev, *Origin of the Human Race*.

31. 'New Face for Kenya Hominid?', *Science* 316(2007): 27.

32. *Homo habilis*, first discovered in Olduvai Gorge in Tanzania; Tattersall and Schwartz, *Extinct Humans*.

33. *Homo erectus*, described originally as *Pithecanthropus erectus*, from fossils found in Java between 1891 and 1898. E. Dubois, 'Pithecanthropus erectus du Pliocene de Java', *P. V. Bull. Soc. Belge Geol.* 9(1895): 151–60; M. H. Day, *Guide to Fossil Man*, 4th edn (London: Cassell, 1986).

34. F. Spoor et al., 'Implications of New Early *Homo* Fossils from Ileret, East of Lake Turkana, Kenya', *Nature* 448(2007): 688–91.

35. J. Kappelman, 'The Evolution of Body Mass and Relative Brain Size in Fossil Hominids', *J. Hum. Evol.* 30(1996): 243–76.

Chapter 2—Once We Were Not Alone

1. The Hobbit was the popular name given to the tiny, small-brained, humans reported to have lived as recently as 18 thousand years ago on the island of Flores in Indonesia. The Hobbit stood a metre tall and had a brain volume of 380 cc. P. Brown et al., 'A New Small-Bodied Hominin from the Late Pleistocene of Flores, Indonesia', *Nature* 431(2004): 1055–61; M. J. Morwood et al., 'Archaeology and Age of a New Hominin from Flores in Eastern Indonesia', *Nature* 431(2004): 1087–91.

2. These proto-humans were first brought to the fore in the mid-1990s, and were named *Homo georgicus* in 2002. They stood 1.45–1.66 metres tall, weighed 40–50 kilos, and had brain volumes in the range 600–780 cc. L. Gabunia and L. Vekua, 'A Plio-Pleistocene Hominid from Dmanisi, East Georgia, Caucasus', *Nature* 373(1995): 509–12; L. Gabunia et al., 'Earliest Pleistocene Hominid Cranial Remains from Dmanisi, Republic of Georgia: Taxonomy, Geological Setting, and Age', *Science* 288(2000): 1019–25; A. Vekua et al., 'A New Skull of Early *Homo* from Dmanisi, Georgia', *Science* 297(2002): 85–9; L. Gabounia et al., 'Découverte d'un nouvel hominidé à Dmanissi (Transcaucasie, Géorgie)', *C. R. Palevol.* 1(2002): 243–53.

3. M. Balter, 'Skeptics Question Whether Flores Hominid Is a New Species', *Science* 306(2004): 1116.

4. M. J. Morwood et al., 'Further Evidence for Small-Bodied Hominins from the Late Pleistocene of Flores, Indonesia', *Nature* 437(2005): 1012–17.

5. D. Argue et al., '*Homo floresiensis*: Microcephalic, Pygmoid, Australopithecus, or Homo?', *J. Human Evol.* 51(2006): 360–74.

6. The skull resembled that of a 1.78-mya *Homo erectus* from East Africa. African *H. erectus* is considered a separate species from Asian erectus by some and classified as *H. ergaster* (Working Man), although I consider them all to belong to a single polytypic '*H. erectus*' species in this book. I. Tattersall and J. Schwartz, *Extinct Humans* (Boulder, CO: Westview Press, 2000); B. Asfaw et al., 'Remains of *Homo erectus* from Bouri, Middle Awash, Ethiopia', *Nature* 416(2002): 317–20.

7. *Australopithecus garhi*: Surprise (in Afar language) Man, a 2.5-mya proto-human from the Middle Awash in Ethiopia. This is the species that may have been responsible for the earliest known tools at Gona, Ethiopia (Ch. 1). B. Asfaw,

'*Australopithecus garhi*: A New Species of Early Hominid from Ethiopia', *Science* 284(1999): 629–35.

8. The age of *A. garhi*.

9. M. W. Tocheri et al., 'The Primitive Wrist of *Homo floresiensis* and Its Implications for Hominin Evolution', *Science* 317(2007): 1743–5.

10. Morwood et al., 'Further Evidence for Small-Bodied Hominins.'

11. A. Brumm et al., 'Early Stone Technology on Flores and Its Implications for *Homo floresiensis*', *Nature* 441(2006): 624–8.

12. D. Lordkipanidze et al., 'The Earliest Toothless Hominin Skull', *Nature* 434(2005): 717–18.

13. Tattersall and Schwartz, *Extinct Humans*; Asfaw et al., 'Remains of *Homo erectus*'; R. Dennell and W. Roebroeks, 'An Asian Perspective on Early Human Dispersal from Africa', *Nature* 438(2005): 1099–104.

14. H. Dowsett et al., 'Joint Investigations of the Middle Pliocene Climate I: PRISM Paleoenvironmental Reconstructions', *Glob. Planet. Change* 9(1994): 169–95.

15. J. H. Cooper, 'First Fossil Record of Azure-Winged Magpie *Cyanopica cyanus* in Europe', *Ibis* 142(2000): 150–1.

16. K. W. Fok et al., 'Inferring the Phylogeny of Disjunct Populations of the Azure-Winged Magpie *Cyanopica cyanus* from Mitochondrial Control Region Sequences', *Proc. R. Soc. Lond. B.* 269(2002): 1671–9; A. Kryukov et al., 'Synchronic East–West Divergence in Azure-Winged Magpies (*Cyanopica cyanus*) and Magpies (*Pica pica*)', *J. Zool. Syst. Evol. Res.* 42(2004): 342–51.

17. J-F. Ghienne et al., 'The Holocene Giant Lake Chad Revealed by Digital Elevation Models', *Quat. Int.* 87(2002): 81–5; K. White and D. Mattingly, 'Ancient Lakes of the Sahara', *Amer. Sci.* 94(2006): 58–66.

18. T. Shine et al., 'Rediscovery of Relict Populations of the Nile Crocodile *Crocodylus niloticus* in south-east Mauritania, with observations on their natural history', *Oryx* 35(2001): 260–2.

19. C. Finlayson, 'Biogeography and Evolution of the Genus *Homo*', *Trends Ecol. Evol.* 20(2005): 457–63.

20. Dennell and Roebroeks, 'An Asian Perspective'.

21. Early sites (2.0–1.5 mya) with tools and no fossils (or fossils that are not diagnostic) include Erk-el-Ahmar and Ubeidiya, Israel; Ain Hanech, Algeria; Niwehan, China; and Riwat, Pakistan. H. Ron and S. Levi, 'When Did Hominids First Leave Africa?: New High-Resolution Magnetostratigraphy from the Erk-el-Ahmar Formation, Israel', *Geology* 29(2001): 887–90; M. Bellmaker et al., 'New Evidence for Hominid Presence in the Lower Pleistocene of the Southern Levant', *J. Hum. Evol.* 43(2002): 43–56; M. Sahnouni et al., 'Further Research at the Oldowan Site of Ain Hanech, North-eastern Algeria', *J. Hum. Evol.* 43(2002): 925–37; R. X. Zhu et al., 'New Evidence on the Earliest Human Presence at High Northern Latitudes in Northeast Asia', *Nature* 431(2004): 559–62; Dennell and Roebroeks, 'An Asian Perspective'.

22. H. T. Bunn, 'Hunting, Power Scavenging, and Butchering by Hadza Foragers and by Plio-Pleistocene *Homo*', in C. B. Stanford and H. T. Bunn (eds), *Meat-Eating and Human Evolution* (Oxford: Oxford University Press, 2001), 199–218.

23. N. Goren-Inbar et al., 'Nuts, Nut Cracking, and Pitted Stones at Gesher Benot Ya'aqov, Israel', *Proc. Natl. Acad. Sci. USA* 99(2002): 2455–60; N. Goren-Inbar et al., *The Acheulian Site of Gesher Benot Ya'aqov, Israel* (Oxford: Oxbow Books, 2002).

24. L. C. Aiello and P. Wheeler, 'The Expensive Tissue Hypothesis: The Brain and Digestive System in Human and Primate Evolution', *Curr. Anthropol.* 36(1995): 199–221; Stanford and Bunn (eds), *Meat-Eating and Human Evolution*.

25. C. B. Stanford, *The Hunting Apes, Meat Eating and the Origins of Human Behavior* (Princeton, NJ: Princeton University Press, 1999); D. P. Watts and J. C. Mitani, 'Hunting Behavior of Chimpanzees at Ngogo, Kibale National Park, Uganda', *Int. J. Primatol.* 23(2002): 1–28.

26. S. C. Strum, 'Baboon Cues for Eating Meat', *J. Hum. Evol.* 12(1983): 327–36; R. J. Rhine et al., 'Insect and Meat Eating among Infant and Adult Baboons (*Papio cynocephalus*) of Mikumi National Park, Tanzania', *Am. J. Phys. Anthropol.* 70(1986): 105–18.

27. J. Sugardjito and N. Nurhuda, 'Meat-Eating Behaviour in Wild Orang utans', *Pongo pygmaeus*, *Primates* 22(1981): 414–16.

28. S. S. Singer et al., 'Molecular Cladistic Markers in New World Monkey Phylogeny (*Platyrrhini, Primates*), *Mol. Phylog. Evol.* 26(2003): 490–501; L. M. Rose, 'Meat and the Early Human Diet', in Stanford and Bunn (eds), *Meat-Eating and Human Evolution*, 141–59. Interestingly Capuchin Monkeys (*Cebus libidinosus*) use anvils and stone pounding tools in the wild to crack open nuts: D. Fragaszy et al., 'Wild Capuchin Monkeys (*Cebus libidinosus*) Use Anvils and Stone Pounding Tools', *Am. J. Primatol.* 64(2004): 359–66.

29. M. Pickford, 'Incisor–Molar Relationships in Chimpanzees and Other Hominoids: Implications for Diet and Phylogeny, *Primates* 46(2005): 21–32.

30. M. Mudelsee and K. Stattegger, 'Exploring the Structure of the Mid-Pleistocene Revolution with Advanced Methods of Time-Series Analysis', *Geol. Rundsch* 86(1997): 499–511.

31. N. J. Shackleton, 'New Data on the Evolution of Pliocene Climatic Variability', in E. S. Vrba et al. (eds), *Paleoclimate and Evolution with Emphasis on Human Origins* (New Haven, CT: Yale University Press, 1995), 242–8; P. B. deMenocal, 'Plio-Pleistocene African Climate', *Science* 270(1995): 53–9.

32. *Homo erectus* fossils from Ngandong and Sambungmacan in Central Java are thought to be morphologically advanced. Dating of fossil bovid teeth collected from the *H. erectus* levels produced mean ages of 27 ± 2 to 53.3 ± 4 thousand years ago. The results are controversial. On the mainland, *H. erectus* survived at least until 300 thousand years ago on Zhoukoudian in China. C. C. Swisher III et al., 'Latest *Homo erectus* of Java: Potential Contemporaneity with *Homo sapiens* in Southeast Asia', *Science* 274(1996): 1870–4; R. Grun et al., 'ESR Analysis

of Teeth from the Palaeoanthropological Site of Zhoukoudian, China', *J. Hum. Evol.* 32(1997): 83–91.

33. J. D. Clark et al., 'African *Homo erectus*: Old Radiometric Ages and Young Oldowan Assemblages in the Middle Awash Valley, Ethiopia', *Science* 264(1994): 1907–10.

34. Fossils (approximately from 600 to 300 thousand years ago) ascribed to *H. heidelbergensis* include, depending on the authority, those from Bodo (Ethiopia), Broken Hill (Zambia), Elandsfontein (South Africa), Lake Ndutu (Tanzania), Petralona (Greece), Arago (France), Bilzingsleben (Germany), Mauer (Germany), Steinheim (Germany), Vertesszöllös (Hungary), Sima de los Huesos (Spain), Swanscombe (United Kingdom), Boxgrove (United Kingdom), Narmada (India), Dali (China), and Jinniushan (China). G. P. Rightmire, 'Patterns of Hominid Evolution and Dispersal in the Middle Pleistocene', *Quat. Int.* 75(2001): 77–84; G. P. Rightmire, 'Human Evolution in the Middle Pleistocene: The Role of *Homo heidelbergensis*', *Evol. Anthropol.* 6(1998): 218–27; A. R. Sankhyan, 'Fossil clavicle of a Middle Pleistocene Hominid from the Central Narmada Valley, India', *J. Hum. Evol.* 32(1997): 3–16; Tattersall and Schwartz, *Extinct Humans*.

35. A Gómez-Olivencia et al., 'Metric and Morphological Study of the Upper Cervical Spine from the Sima de los Huesos Site (Sierra de Atapuerca, Burgos, Spain)', *J. Hum. Evol.* 53(2007): 6–25.

36. Rightmire, 'Patterns of Hominid Evolution'.

37. *Homo antecessor*. J. M. Bermúdez de Castro et al., 'A Hominid from the Lower Pleistocene of Atapuerca, Spain: Possible Ancestor to Neanderthals and Modern Humans', *Science* 276(1997): 1392–5; E. Carbonell et al., 'The First Hominin of Europe', *Nature* 452(2008): 465–9.

38. J. M. Bermúdez de Castro et al., 'Gran Dolina-TD6 versus Sima de los Huesos Dental Samples from Atapuerca: Evidence of Discontinuity in the European Pleistocene Population?', *J. Archaeol. Sci.* 30(2003): 1421–8.

Chapter 3—Failed Experiments

1. W. Davies and R. Charles (eds), *Dorothy Garrod and the Progress of the Palaeolithic: Studies in the Prehistoric Archaeology of the Near East and Europe* (Oxford: Oxbow Books, 1999).

2. A. Keith, 'Mount Carmel Man: His Bearing on the Ancestry of Modern Races', in G. G. MacCurdy (ed.), *Early Man* (New York: Lippincot, 1937); T. D. McCown and A. Keith, *The Stone Age of Mt. Carmel*, Vol. 2: *The Fossil Human Remains from the Levalloiso-Mousterian* (Oxford: Clarendon Press, 1939), 41–52.

3. R. Grün et al., 'U-series and ESR Analyses of Bones and Teeth Relating to the Human Burials from Skhul', *J. Hum. Evol.* 49(2005): 316–34.

4. E. Tchernov, 'The Faunal Sequence of the Southwest Asian Middle Paleolithic in Relation to Hominid Dispersal Events', in T. Akazawa, K. Aochi, and O. Bar-Yosef (eds), *Neandertals and Modern Humans in Western Asia* (New York: Plenum Press, 1998), 77–90.

5. The climate of the Middle East has been highly variable over the past 400 thousand years. The variations are the result of the relative positions of the high-latitude north-eastern Atlantic/Mediterranean front system and the low-latitude African/west Asian monsoon systems. The overall trend was for very humid and rainy conditions during warm interglacials and cool and dry conditions during glacial maxima and cold Heinrich events. In between these extreme conditions, dry and warm periods affected the region but cool and humid ones were more localized. The detailed picture has been constructed from marine cores and well-dated cave speleothem records: A. Almogi-Labin, M. Bar-Matthews, and A. Ayalon, 'Climate Variability in the Levant and North-east Africa during the Late Quaternary Based on Marine and Land Records', in N. Goren-Inbar and J. D. Speth (eds), *Human Peleoecology in the Levantine Corridor* (Oxford: Oxbow Books, 2004), 117–34. See also A. Brauer et al., 'Evidence for Last Interglacial Chronology and Environmental Change from Southern Europe', *Proc. Natl. Acad. Sci. USA* 104(2007): 450–5, for a detailed record of interglacial conditions in the Mediterranean.

6. Grün et al., 'U-series and ESR Analyses of Bones and Teeth'. The Tabûn C1 Neanderthal has been dated to 122 ± 16 thousand years ago; R. Grün and C. Stringer, 'Tabun Revisited: Revised ESR Chronology and New ESR and U-series Analyses of Dental Material from Tabun C1', *J. Hum. Evol.* 39(2000): 601–12.

7. J. J. Shea, 'The Middle Paleolithic of the East Mediterranean Levant', *J. World Prehist.* 17(2003): 313–94.

8. Tchernov, 'The Faunal Sequence'.

9. FAUNMAP Working Group, 'Spatial Response of Mammals to Late Quaternary Environmental Fluctuations', *Science* 272(1996): 1601–6.

10. 'Palaearctic' is the zoogeographical region that encompasses Europe, northern Asia, and North Africa.

11. The Middle East is close to the northern edge of a belt that runs from West Africa to China and south to South Africa, in which water is the main limiting factor to net primary productivity. The Iberian Peninsula, Australia, and areas of North and South America also fall under this regime. Temperature and solar radiation are not key limiting factors in these regions. Rain forest areas of Africa, South America, and south-east Asia are outside of this regime. G. Churkina and S. W. Running, 'Contrasting Climatic Controls on the Estimated Productivity of Global Terrestrial Biomes', *Ecosystems* 1(1998): 206–15.

12. Tchernov, 'The Faunal Sequence'; O. Bar-Yosef, 'The Middle and Early Upper Paleolithic in Southwest Asia and Neighboring Regions', in O. Bar-Yosef and D. Pilbeam (eds), *The Geography of Neanderthals and Modern Humans in Europe*

and the Greater Mediterranean, Peabody Museum Bulletin 8 (Cambridge, MA: Harvard University Press, 2000), 107–56.

13. J. Clutton-Brock, *A Natural History of Domesticated Mammals* (London: Natural History Museum, 1999).

14. *Equus tabeti*, see Tchernov, 'The Faunal Sequence'.

15. Clutton-Brock, *Natural History of Domesticated Mammals*.

16. S. Cramp (ed.), *Handbook of the Birds of Europe the Middle East and North Africa, The Birds of the Western Palearctic*, Vol. 1: *Ostrich to Ducks* (Oxford: Oxford University Press, 1977).

17. Tchernov, 'The Faunal Sequence'.

18. The roe deer, *Capreolus capreolus*, a species of dense forest, is only present in very small numbers.

19. Bar-Yosef, 'The Middle and Early Upper Paleolithic'.

20. ESR dating of dental material gave an age of 122 ± 16 thousand years ago. The Tabûn C1 Neanderthal is thought to have been a burial from Layer B to Layer C so the most likely contemporaneous fauna is that of Layer B. Grün and Stringer, 'Tabun revisited'.

21. Phytoliths, literally 'plant stones', are rigid, microscopic bodies found in many kinds of plants and recoverable from archaeological sites. Phytoliths recovered from the Tabûn layer associated with the 122-thousand-year-old Neanderthal were of the kind found in present-day Mediterranean woodland vegetation of the area. Although the precise allocation of particular phytolith types to plant species could not be confirmed, the main plant species with phytoliths similar to those recovered included the evergreen Palestine oak (*Quercus calliprinos*), the deciduous Tabor oak (*Quercus ithaburensis*), carob (*Ceratonia siliqua*), and olive (*Olea europaea*). All these species grow in the area today. R. M. Albert et al., 'Mode of Occupation of Tabun Cave, Mt Carmel, Israel during the Mousterian Period: A Study of the Sediments and Phytoliths', *J. Arch. Sci.* 26(1999): 1249–60.

22. At Nzalet Khater, Egypt, dated to 37, 570 + 350, −310 years. E. Trinkaus, 'Early Modern Humans', *Ann. Rev. Anthropol.* 34(2005): 207–30.

23. H. Valladas et al., 'Thermoluminescence Dates for the Neanderthal Burial Site at Kebara in Israel', *Nature* 330(1987): 159–60; H. P. Schwarcz et al., 'ESR Dating of the Neanderthal Site, Kebara Cave, Israel', *J. Archaeol. Sci.* 16(1989): 653–9; H. Valladas et al., 'TL Dates for the Neanderthal Site of the Amud Cave, Israel', *J. Archaeol. Sci.* 26(1999): 259–68; W. J. Rink et al., 'Electron Spin Resonance (ESR) and Thermal Ionization Mass Spectrometric (TIMS) ^{230}Th/^{234}U Dating of Teeth in Middle Paleolithic Layers at Amud Cave, Israel', *Geoarchaeology* 16(2001): 701–17.

24. Almogi-Labin, Bar-Matthews, and Ayalon, 'Climate Variability'.

25. The Middle Palaeolithic technology known as Levantine Mousterian.

26. The sub-Saharan technologies are generally ascribed to the Middle Stone Age that approximately corresponds to the Eurasian Middle Palaeolithic. The

African Middle Stone Age resembles the Levantine Mousterian by the regular presence of stone spear points which may be linked to ambush hunting of large mammals. J. J. Shea, 'Neandertals, Competition, and the Origin of Modern Human Behavior in the Levant', *Evol. Anthropol.* 12(2003): 173–87.

27. Trinkaus, 'Early Modern Humans'.

28. P. Mellars and C. Stringer (eds), *The Human Revolution: Behavioural and Biological Perspectives in the Origins of Modern Humans* (Edinburgh: Edinburgh University Press, 1989).

29. R. G. Klein, 'Archeology and the Evolution of Human Behavior', *Evol. Anthropol.* 9(2000): 17–36.

30. C. Finlayson, 'Biogeography and Evolution of the genus *Homo*',*Trends Ecol. Evol.* 20(2005): 457–63.

31. C. Henshilwood et al., 'Middle Stone Age Shell Beads from South Africa', *Science* 304(2004): 404.

32. Ten, possibly 12, pieces of worked ochre pigment out of 57 pieces of ochre found at Pinnacle Point, South Africa, and dated to 164 ± 12 thousand years ago. C. Marean et al., 'Early Human Use of Marine Resources and Pigment in South Africa during the Middle Pleistocene', *Nature* 449(2007): 905–9.

33. Henshilwood et al., 'Middle Stone Age Shell Beads'; M. Vanhaeren et al., 'Middle Paleolithic Shell Beads in Israel and Algeria', *Science* 312(2006): 1785–8; A. Bouzouggar et al., '82,000-Year-Old Shell Beads from North Africa and Implications for the Origins of Modern Human Behaviour', *Proc. Natl. Acad. Sci. USA* 104(2007): 9964–9.

34. Dates are optically stimulated luminescence (osl) dates. Thermoluminescence dates on burnt flints in the same level of Blombos Cave gave estimated ages of 77 ± 6 thousand years ago. Henshilwood et al., 'Middle Stone Age Shell Beads'.

35. F. E. Grine and C. S. Henshilwood, 'Additional Human Remains from Blombos Cave, South Africa (1999–2000 Excavations)', *J. Hum. Evol.* 42(2002): 293–302.

36. M. Vanhaeren, 'Middle Paleolithic Shell Beads'.

37. Ibid.

38. Bouzouggar et al., '82,000-Year-Old Shell Beads'.

39. On the basis of 23 individual molluscs belonging to 10 species, assumed to have been consumed, Marean, 'Early Human Use of Marine Resources'. Two human fossils have been recovered from this site but they are undiagnostic; C. Marean et al., 'Paleoanthropological Investigations of Middle Stone Age Sites at Pinnacle Point, Mossel Bay (South Africa): Archaeology and Hominid Remains from the 2000 Field Season', *PaleoAnthropology* 5(2004): 14–83.

40. C. Stringer, 'Coasting out of Africa', *Nature* 405(2000): 24–7.

41. M. Pagani et al., 'Marked Decline in Atmospheric Carbon Dioxide Concentrations during the Paleogene', *Science* 309(2006): 600–3.

42. T. E. Cerling et al., 'Global Vegetation Change through the Miocene/Pliocene Boundary', *Nature* 389(1997): 153–8.
43. L. Ségalen et al., 'Timing of C_4 Grass Expansion across Sub-Saharan Africa', *J. Hum. Evol.* 53(2007): 549–59.
44. S. F. Greb et al., 'Evolution and Importance of Wetlands in Earth History', *Geol. Soc. Amer., Special Paper* 399(2006): 1–40; G. P. Nicholas, 'Wetlands and Hunter-Gatherers: A Global Perspective', *Curr. Anthropol.* 39(1998): 720–31; C. Finlayson, *Neanderthals and Modern Humans: An Ecological and Evolutionary Perspective* (Cambridge: Cambridge University Press, 2004).
45. R. Dennell, 'Dispersal and Colonisation, Long and Short Chronologies: How Continuous Is the Early Pleistocene Record for Hominids Outside East Africa?', *J. Hum. Evol.* 45(2003): 421–40.
46. D. M. Bramble and D. E. Lieberman, 'Endurance Running and the Evolution of *Homo*', *Nature* 432(2004): 345–52; K. L. Steudel-Numbers et al., 'The Effect of Lower Limb Length on the Energetic Cost of Locomotion: Implications for Fossil Hominins', *J. Hum. Evol.* 47(2004): 95–109; K. L. Steudel-Numbers, 'Energetics in *Homo erectus* and Other Early Hominins: The Consequences of Increased Lower-Limb Length', *J. Hum. Evol.* 51(2006): 445–53; K. L. Steudel-Numbers et al., 'The Evolution of Human Running: Effects of Changes in Lower-Limb Length on Locomotor Economy', *J. Hum. Evol.* 53(2007): 191–6.
47. Churkina and Running, 'Contrasting Climatic Controls'.
48. 195 ± 5 thousand years ago in Omo Kibish, Ethiopia; 160–154 thousand years ago in Herto, Ethiopia; 133 ± 2 thousand years ago in Singa, Sudan. I. MacDougall et al., 'Stratigraphic Placement and Age of Modern Humans from Kibish, Ethiopia', *Nature* 433(2005): 733–6; J. D. Clark et al., 'Stratigraphic, Chronological and Behavioural Contexts of Pleistocene Homo sapiens from Middle Awash, Ethiopia', *Nature* 423(2003): 747–52; F. McDermott et al., 'New Late-Pleistocene Uranium–Thorium and ESR Dates for the Singa Hominid (Sudan)', *J. Hum. Evol.* 31(1996): 507–16.
49. Fossils from Jebel Irhoud, Morocco, are classified as archaic modern human (proto-Ancestors in our terminology). One specimen has been dated by direct uranium series/electron spin resonance to 160 ± 16 thousand years ago. J-J. Hublin, 'Modern-Nonmodern Hominid Interactions: A Mediterranean Perspective', in Bar-Yosef and Pilbeam (eds), *Geography of Neanderthals*, 157–82; T. M. Smith et al., 'Earliest Evidence of Modern Human Life History in North African Early *Homo sapiens*', *Proc. Natl. Acad. Sci. USA* 104(2007): 6128–33.
50. A. Ayalon et al., 'Climatic Conditions during Marine Oxygen Isotope Stage 6 in the Eastern Mediterranean Region from the Isotopic Composition of Speleothems of Soreq Cave, Israel', *Geology* 30(2002): 303–6.
51. S. Oppenheimer, *Out of Eden: The Peopling of the World* (London: Robinson, 2004).
52. S. Wells, *The Journey of Man: A Genetic Odyssey* (London: Penguin, 2002).
53. Oppenheimer, *Out of Eden*; Wells, *The Journey of Man*.

Chapter 4—Stick to What You Know Best

1. G. H. Orians and J. H. Heerwagen, 'Evolved Responses to Landscapes', in J. H. Barkow et al. (eds), *The Adapted Mind: Evolutionary Psychology and the Generation of Culture* (New York: Oxford University Press, 1992); G. H. Orians, 'Human Behavioural Ecology: 140 Years without Darwin Is Too Long', *Bull. Ecol. Soc. Amer.* 79(1998): 15–28; *idem,* 'Aesthetic Factors', *Encyclopaedia of Biodiversity* 1(2001): 45–54.

2. Z. Majid, 'The West Mouth, Niah, in the Prehistory of Southeast Asia', *Sarawak Mus. J.* 23(1982): 1–200.

3. T. Harrisson, 'Radio Carbon C-14 Datings from Niah: A Note', *Sarawak Mus. J.* 9(1959): 136–8.

4. G. Barker et al., 'The "Human Revolution" in Lowland Tropical Southeast Asia: The Antiquity and Behavior of Anatomically Modern Humans at Niah Cave (Sarawak, Borneo)', *J. Hum. Evol.* 52(2007): 243–61.

5. Unless specifically stated otherwise, radiocarbon dates are given in uncalibrated form. The carbon content of the atmosphere has varied through time, which means that radiocarbon years may not be equivalent to calendar years and require calibration. Because there is no reliable calibration curve for radiocarbon ages older than 26 thousand years ago, it is safer to use uncalibrated dates for dates older than 26 thousand years ago. Calibrated dates are generally older than uncalibrated ones but the difference varies with the atmospheric carbon content at any given time. P. Reimer et al., 'Comment on "Radiocarbon Calibration Curve Spanning 0 to 50,000 Years BP Based on Paired ^{230}Th/^{234}U/^{238}U and ^{14}C Dates on Pristine Corals" by R. G. Fairbanks et al.', *Quat. Sci. Rev.* 24(2005): 1781–96.

6. C. O. Hunt et al., 'Modern Humans in Sarawak, Malaysian Borneo, during Oxygen Isotope Stage 3: Palaeoenvironmental Evidence from the Great Cave of Niah', *J. Arch. Sci.* 34(2007): 1953–69.

7. Niah is 3° north of the equator. At the height of the last Ice Age air temperatures in Borneo were down by 6–7 °C and rainfall was reduced by 30–50%; Hunt et al., 'Modern Humans in Sarawak'; M. I. Bird et al., 'Palaeoenvironments of Insular Southeast Asia during the Last Glacial Period: A Savanna Corridor in Sundaland?', *Quat. Sci. Rev.* 24(2005): 2228–42.

8. Barker et al., 'The "Human Revolution" '.

9. In the 100- to 1000-kilogramme-size category; C. Finlayson, *Neanderthals and Modern Humans: An Ecological and Evolutionary Perspective* (Cambridge: Cambridge University Press, 2004).

10. Barker et al., 'The "Human Revolution" '.

11. Ibid.

12. L. Beaufort et al., 'Biomass Burning and Oceanic Primary Production Estimates in the Sulu Sea Area over the Last 380 kyr and the East Asian Monsoon Dynamics', *Mar. Geol.* 201(2003): 53–65; G. Anshari et al., 'Environmental

Change and Peatland Forest Dynamics in the Lake Sentarum Area, West Kalimantan, Indonesia', *J. Quat. Sci.* 19(2004): 637–55.

13. P. A. Underhill et al., 'The Phylogeography of Y Chromosome Binary Haplotypes and the Origins of Modern Human Populations', *Ann. Hum. Genet.* 65(2001): 43–62; S. Wells, *The Journey of Man: A Genetic Odyssey* (London: Penguin, 2002); S. Oppenheimer, *Out of Eden: The Peopling of the World* (London: Robinson, 2004); S. Barnabas et al., 'High-Resolution mtDNA Studies of the Indian Population: Implications for Palaeolithic Settlement of the Indian Subcontinent', *Ann. Hum. Genet.* 70(2005): 42–58; V. Macaulay et al. 'Single, Rapid Coastal Settlement of Asia Revealed by Analysis of Complete Mitochondrial Genomes', *Science* 308(2005): 1034–6.

14. B. J. Szabo et al., 'Ages of Quaternary Pluvial Episodes Determined by Uranium-Series and Radiocarbon Dating of Lacustrine Deposits of Eastern Sahara', *Palaeogeogr., Palaeoclimatol., Palaeoecol.* 113(1995): 227–42; D. Fleitmann et al., 'Changing Moisture Sources over the Last 330,000 Years in Northern Oman from Fluid-Inclusion Evidence in Speleothems', *Quat. Res.* 60(2003): 223–32; J. K. Osmond and A. A. Dabous, 'Timing and Intensity of Groundwater Movement during Egyptian Sahara Pluvial Periods by U-series Analysis of Secondary U in Ores and Carbonates', *Quat. Res.* 61(2004): 85–94; J. R. Smith et al., 'A Reconstruction of Quaternary Pluvial Environments and Human Occupations Using Stratigraphy and Geochronology of Fossil-Spring Tufas, Kharga Oasis, Egypt', *Geoarchaeol.* 19(2004): 407–39; A. Vaks et al., 'Desert Speleothems Reveal Climatic Window for African Exodus of Early Modern Humans', *Geology* 35(2007): 831–4.

15. M. M. Lahr and R. Foley, 'Multiple Dispersals and Modern Human Origins', *Evol. Anthropol.* 3(1994): 48–60.

16. L. Quintana-Murci et al., 'Genetic Evidence of an Early Exit of *Homo sapiens sapiens* through Eastern Africa', *Nat. Genet.* 23(1999): 437–41.

17. R. C. Walter et al., 'Early Human Occupation of the Red Sea Coast of Eritrea during the Last Interglacial', *Nature* 405(2000): 65–9. Further evidence was claimed in 2008 of the exploitation of giant clams in the Red Sea starting at this time but the evidence is ambiguous: C. Richter, et al., 'Collapse of a New Living Species of Giant Clam in the Red Sea', *Curr. Biol.* 18(2008): 1–6.

18. C. Marean et al., 'Early Human Use of Marine Resources and Pigment in South Africa during the Middle Pleistocene', *Nature* 449(2007): 905–9.

19. J. H. Bruggemann et al., 'Stratigraphy, Palaeoenvironments and Model for the Deposition of the Abdur Reef Limestone: Context for an Important Archaeological Site from the Last Interglacial on the Red Sea Coast of Eritrea', *Palaeogeogr., Palaeoclimatol., Palaeoecol.* 203(2004): 179–206.

20. E. J. Rohling et al., 'High Rates of Sea-Level Rise during the Last Interglacial Period', *Nat. Geosc.* 1(2007): 38–42.

21. A. Carpenter, 'Monkeys Opening Oysters', *Nature* 36(1887): 53.

22. S. Malaivijitnond et al., 'Stone-Tool Usage by Thai Long-Tailed Macaques (*Macaca fascicularis*)', *Am. J. Primatol.* 69(2007): 227–33.
23. G. V. Glazko and M. Nei, 'Estimation of Divergence Times for Major Lineages of Primate Species', *Mol. Biol. Evol.* 20(2003): 424–34.
24. A. Brumm et al., 'Early Stone Technology on Flores and Its Implications for *Homo floresiensis*', *Nature* 441(2006): 624–8.
25. C. Abegg and B. Thierry, 'Macaque Evolution and Dispersal in Insular South-East Asia', *Biol. J. Linn. Soc.* 75(2002): 555–76.
26. R. G. Klein, *The Human Career: Human Biological and Cultural Origins* (Chicago: Chicago University Press, 1999).
27. C. A. Fernandes et al., 'Absence of Post-Miocene Red Sea Land Bridges: Biogeographic Implications', *J. Biogeogr.* 33(2006): 961–6.
28. J. S. Field and M. M. Lahr, 'Assessment of the Southern Dispersal: GIS-Based Analyses of Potential Routes at Oxygen Isotopic Stage 4', *J. World Prehist.* 19(2005): 1–45.
29. P. Van Peer, 'The Nile Corridor and the Out-of-Africa Model', *Curr. Anthropol.* 39(suppl.) (1998): S115–40.
30. Szabo et al., 'Ages of Quaternary Pluvial Episodes'; Fleitmann et al., 'Changing Moisture Sources'; Osmond and Dabous, 'Timing and Intensity of Groundwater Movement'; Smith et al., 'A Reconstruction of Quaternary Pluvial Environments'; Vaks et al., 'Desert Speleothems Reveal Climatic Window'.
31. Van Peer, 'The Nile Corridor'.
32. In the absence of fossils, we cannot be certain of the identity of these people. Given the apparent absence of Neanderthals anywhere south of the Levant, it is assumed that they are proto-Ancestors or Ancestors, or both. Perhaps one group was related to the proto-Ancestors of Skhul and Qafzeh (Ch. 3).
33. Van Peer, 'The Nile Corridor'.
34. M. D. Petraglia and A. Alsharekh, 'The Middle Palaeolithic of Arabia: Implications for Modern Human Origins, Behaviour and Dispersals', *Antiquity* 77(2003): 671–84.
35. Ibid.
36. D. Schmitt and S. E. Churchill, 'Experimental Evidence Concerning Spear Use in Neandertals and Early Modern Humans', *J. Archaeol. Sci.* 30(2003): 103–14.
37. M. Cremaschi et al., 'Some Insights on the Aterian in the Libyan Sahara: Chronology, Environment, and Archaeology', *Afr. Archaeol. Rev.* 15(1998): 261–86; A. Debénath, 'Le peuplement prEhistorique du Maroc: données récentes et problemes', *L'Anthropol.* 104(2000): 131–45; A. Bouzouggar et al., 'Étude des ensembles lithiques atériens de la grotte d'El Aliya à Tanger (Maroc)', *L'Anthropol.* 106(2002): 207–48; A. C. Haour, 'One Hundred Years of Archaeology in Niger', *J. World Prehist.* 17(2003): 181–234; E. A. A. Garcea, 'Crossing Deserts and Avoiding Seas: Aterian North African-European Relations', *J. Anthropol. Res.* 60(2004): 27–53; B. E. Barich et al., 'Between the Mediterranean and the Sahara: Geoarchaeological Reconnaissance in the Jebel Gharbi,

Libya', *Antiquity* 80(2006): 567–82; A. Bouzouggar et al., '82,000-Year-Old Shell Beads from North Africa and Implications for the Origins of Modern Human Behaviour', *Proc. Natl. Acad. Sci. USA* 104(2007): 9964–9; N. Mercier et al., 'The Rhafas Cave (Morocco): Chronology of the Mousterian and Aterian Archaeological Occupations and Their Implications for Quaternary Geochronology Based on Luminescence (TL/OSL) Age Determinations', *Quat. Geochronol.* 2(2007): 309–313.

38. D. Geraads, 'Faunal Environment and Climatic Change in the Middle/Late Pleistocene of North-western Africa', Abstracts of *Modern Origins: A North African Perspective* (Leipzig: Max Planck, 2007).

39. J.-J. Hublin et al., 'Dental Evidence from the Aterian Human Populations of Morocco', ibid.

40. Van Peer, 'The Nile Corridor'.

41. H. V. A. James and M. D. Petraglia, 'Modern Human Origins and the Evolution of Behavior in the Later Pleistocene Record of South Asia', *Curr. Anthropol.* 46(suppl.)(2005): S3–27.

42. Barnabas, 'High-Resolution mtDNA Studies'.

43. K. O. Pope and J. E. Terrell, 'Environmental Setting of Human Migrations in the Circum-Pacific Region', *J. Biogeogr.* 35(2008): 1–21.

44. J. S. Field et al., 'The Southern Dispersal Hypothesis and the South Asian Archaeological Record: Examination of Dispersal Routes through GIS Analysis', *J. Anthropol. Archaeol.* 26(2007): 88–108.

45. A period of very low population size which reduced the human genetic diversity. H. C. Harpending et al., 'The Genetic Structure of Ancient Human Populations', *Curr. Anthropol.* 34(1993): 483–96.

46. S. H. Ambrose, 'Late Pleistocene Human Population Bottlenecks, Volcanic Winter, and Differentiation of Modern Humans', *J. Hum. Evol.* 34(1998): 623–51.

47. Pope and Terrell, 'Environmental Setting of Human Migrations'.

48. Bird, 'Palaeoenvironments of Insular Southeast Asia'.

49. Barker et al., 'The "Human Revolution"'.

50. S. S. Barik et al., 'Detailed mtDNA Genotypes Permit a Reassessment of the Settlement and Population Structure of the Andaman Islands', *Am. J. Phys. Anthropol.* 136(2008): 19–27.

51. G. Hudjashov et al., 'Revealing the Prehistoric Settlement of Australia by Y Chromosome and mtDNA analysis', *Proc. Natl. Acad. Sci. USA* 104(2007): 8726–30.

52. P. Clarke, *Where the Ancestors Walked* (Crow's Nest, NSW: Allen and Unwin, 2003); J. Flood, *Archaeology of the Dreamtime: The Story of Prehistoric Australia and Its People* (Marleston, South Australia: JB Publishing, 2004).

53. J. Bowler et al., 'New Ages for Human Occupation and Climatic Change at Lake Mungo, Australia', *Nature* 421(2003): 837–40.

Chapter 5—Being in the Right Place at the Right Time

1. An adult male was estimated to have weighed just 43.2 kg and a female 28.7 kg; L. R. Berger et al., 'Small-Bodied Humans from Palau, Micronesia', *PLoS One* 3(2008/): e1780, doi:10.1371/journal.pone.0001780.

2. C. Finlayson, *Neanderthals and Modern Humans: An Ecological and Evolutionary Perspective* (Cambridge: Cambridge University Press, 2004).

3. The first Neanderthal fossil DNA sequences were reported in 1997: M. Krings et al., 'Neandertal DNA Sequences and the Origin of Modern Humans', *Cell* 90(1997): 19–30.

4. R. E. Green et al., 'Analysis of One Million Base Pairs of Neanderthal DNA', *Nature* 444(2006): 330–6; J. P. Noonan et al., 'Sequencing and Analysis of Neanderthal Genomic DNA', *Science* 314(2006): 1113–18.

5. C. Lalueza-Fox et al., 'A Melanocortin 1 Receptor Allele Suggests Varying Pigmentation among Neanderthals', *Science* 318(2007): 1453–5.

6. J. Krause et al., 'The Derived FOXP2 Variant of Modern Humans Was Shared with Neandertals', *Curr. Biol.* 17(2007): 1908–12.

7. C. Finlayson, 'Biogeography and Evolution of the Genus *Homo*', *Trends Ecol. Evol.* 20(2005): 457–63.

8. J. D. Wall and S. K. Kim, 'Inconsistencies in Neanderthal Genomic DNA Sequences', *PLoS Genetics* 3(2007): e175; T. D. Weaver et al., 'Close Correspondence between Quantitative- and Molecular-Genetic Divergence Times for Neandertals and Modern Humans', *Proc. Natl. Acad. Sci. USA* 105(2008): 4645–9.

9. I. V. Ovchinnikov et al., 'Molecular Analysis of Neanderthal DNA from the Northern Caucasus', *Nature* 404(2000): 490–3; P. Beerli and S. V. Edwards, 'When Did Neanderthals and Modern Humans Diverge?', *Evol. Anthropol. Suppl.* 1(2002): 60–3.

10. See, for example, C. Stringer and C. Gamble, *In Search of the Neanderthals: Solving the Puzzle of Human Origins* (London: Thames and Hudson, 1993).

11. C. Finlayson and J. S. Carrión, 'Rapid Ecological Turnover and Its Impact on Neanderthal and Other Human Populations', *Trends Ecol. Evol.* 22(2007): 213–22; J. Krause et al., 'Neanderthals in Central Asia and Siberia', *Nature* 449(2007): 902–4.

12. It is not easy to be precise as to the time when, through a process of continuous evolution, the Middle Pleistocene humans of Eurasia could be defined as Heidelberg or Neanderthal Man. Specimens recognizable as Neanderthal appear some time between 200 and 125 thousand years ago.

13. W. J. Burroughs, *Climate Change in Prehistory: The End of the Reign of Chaos* (Cambridge: Cambridge University Press, 2005).

14. W. Roebroeks, N. J. Conard, and T. van Kolfschoten, 'Dense Forests, Cold Steppes, and the Palaeolithic Settlement of Europe', *Curr. Anthropol.* 33(1992): 551–86.

15. See Ch. 2. Where we place the African fossils of *H. heidelbergensis* is a matter of conjecture and depends on when the African and Eurasian lineages became isolated from each other. If this separation took place early in the Middle Pleistocene, as some genetic estimates suggest, then the African form should be given a different name from the Eurasian one, probably Rhodesian Man, *Homo rhodesiensis*. *Homo heidelbergensis* would then be a name exclusive to the Eurasian population that was the predecessor of the Neanderthals but not the Ancestors. A recent study that combines anatomy and genetics strongly suggests a close link between *H. heidelbergensis* and the Neanderthals with modern humans being a separate evolutionary lineage; R. González-José et al., 'Cladistic Analysis of Continuous Modularized Traits Provides Phylogenetic Signals in *Homo* Evolution', *Nature* 453(2008): 775–8.

16. P. deMenocal, 'Plio-Pleistocene African Climate', *Science* 270(1995): 53–9.

17. Southern elephant, *Mammuthus meridionalis*; steppe mammoth, *Mammuthus trogontheri*; woolly mammoth, *Mammuthus primigenius*. A. M. Lister and A. V. Sher, 'The Origin and Evolution of the Woolly Mammoth', *Science* 294(2001): 1094–7.

18. S. L. Vartanyan, V. E. Garrut, and A. V. Sher, 'Holocene Dwarf Mammoths from Wrangel Island in the Siberian Arctic', *Nature* 382(1993): 337–40.

19. Lister and Sher, 'Origin and Evolution of the Woolly Mammoth'.

20. R. G. Klein, *The Human Career: Human Biological and Cultural Origins* (Chicago: Chicago University Press, 1999).

21. T. van Kolfschoten, 'The Eemian Mammal Fauna of Central Europe', *Neth. J. Geosci.* 79(2000): 269–81; D. Pushkina, 'The Pleistocene Easternmost Distribution in Eurasia of the Species Associated with the Eemian *Palaeoloxodon antiquus* Assemblage', *Mammal. Rev.* 37(2007): 224–45.

22. Not all warm periods were wet and not all cold periods were dry but most of them fitted these broad categories.

23. J-C Svenning, 'A Review of Natural Vegetation Openness in North-western Europe', *Biol. Cons.* 104(2002): 133–48.

24. *Tapirus arvernensis*; C. Guérin and M. Patou-Mathis, *Les grands mammifères plio-pleistocenes d'Europe* (Paris: Masson, 1997).

25. R. E. Bodmer, 'Ungulate Frugivores and the Browser-Grazer Continuum', *Oikos* 57(1990): 319–25.

26. A. J. Stuart, 'Mammalian Extinctions in the Late Pleistocene of Northern Eurasia and North America', *Biol. Rev. Camb. Philos. Soc.* 66(1991): 453–562.

27. Two species of hippopotamus lived in Europe in the Pleistocene. One was a European species, *Hippopotamus major* (also known as *Hippopotamus antiquus*). The African *Hippopotamus amphibius*, the same species that survives in Africa today, was the one that survived until the last interglacial in Europe; Pushkina, 'The Pleistocene Easternmost Distribution'.

28. *Bubalus murrensis*.

29. This macaque is also popularly known as the rock ape from its presence on the Rock of Gibraltar and its tailless appearance. Barbary macaques were introduced on Gibraltar by the British in the eighteenth century and went feral.

30. Guérin and Patou-Mathis, *Les grands mammifères plio-pleistocenes*.

31. The narrow-nosed rhinoceros, *Stephanorhinus hemitoechus*, was the smaller of the two species and inhabited wooded steppe; the larger Merck's rhinoceros, *Stephanorhinus kirchbergensis*, lived in woodland. The straight-tusked elephant, *Elephas (Palaeoloxodon) antiquus*, also lived in broadleaf woodland. These species went extinct after the last interglacial but before the last glaciation, some time between 50 and 25 thousand years ago.

32. Stuart, 'Mammalian Extinctions'; Pushkina, 'The Pleistocene Easternmost Distribution'.

33. Finlayson and Carrión, 'Rapid Ecological Turnover'; Krause et al., 'Neanderthals in Central Asia'.

34. M. Pitts and M. Roberts, *Fairweather Eden* (London: Century, 1997); C. Gamble, *The Palaeolithic Societies of Europe* (Cambridge: Cambridge University Press, 1999); S. A. Parfitt et al., 'The Earliest Record of Human Activity in Northern Europe', *Nature* 438(2005): 1008–12.

35. H. Thieme, 'Lower Palaeolithic Hunting Spears from Germany', *Nature* 385(1997): 807–10.

36. 'Classic Neanderthals' is the term often used to describe the fossils readily identifiable as Neanderthal on account of having the full set of anatomical characteristics attributable to them.

37. Finlayson, *Neanderthals and Modern Humans*.

38. A. J. Stuart, 'The Failure of Evolution: Late Quaternary Mammalian Extinctions in the Holarctic', *Quat. Int.* 19(1993): 101–7.

39. S. E. Churchill, 'Of Assegais and Bayonets: Reconstructing Prehistoric Spear Use', *Evol. Anthropol.* 11(2002): 185–6.

40. Klein, *The Human Career*.

41. T. D. Berger and E. Trinkaus, 'Patterns of Trauma among the Neandertals', *J. Archaeol. Sci.* 22(1995): 841–52.

42. Finlayson, *Neanderthals and Modern Humans*.

43. J. R. M. Allen et al., 'Rapid Environmental Changes in Southern Europe during the Last Glacial Period', *Nature* 400(1999): 740–3.

44. Comparative analysis of Neanderthal DNA from various regions of Europe indicates that there was isolation of different populations in refugia during glacial periods; C. Lalueza-Fox et al., 'Mitochondrial DNA of an Iberian Neandertal Suggests a Population Affinity with Other European Neandertals', *Curr. Biol.* 16(2006): R629–30.

45. Finlayson amd Carrión, 'Rapid Ecological Turnover'.

46. The last population of Neanderthals lived in Gorham's Cave between 28 and 24 thousand years ago, several thousand years later than those of northern Iberia and south-western France and at least one thousand years after those

in the Balkans, Crimea, and the Caucasus; C. Finlayson et al., 'Late Survival of Neanderthals at the Southernmost Extreme of Europe', *Nature* 443(2006): 850–3.

Chapter 6—If Only . . .

1. Humans with a slim, gracile, body that had evolved as a way of losing heat effectively in the warm tropical African climates in which they had evolved, e.g. C. Stringer and C. Gamble, *In Search of the Neanderthals: Solving the Puzzle of Human Origins* (London: Thames and Hudson, 1993); R. G. Klein, *The Human Career: Human Biological and Cultural Origins* (Chicago: Chicago University Press, 1999).
2. F. Weindereich, 'The "Neanderthal Man" and the Ancestors of "Homo sapiens" ', *Amer. Anthropol.* 42(1943): 375–83.
3. P. Pettitt, 'Odd Man Out: Neanderthals and Modern Humans', *Brit. Archaeol.* 51(2000): 1–5.
4. F. C. Howell, 'The Evolutionary Significance of Variations and Varieties of "Neanderthal" Man', *Quat. Rev. Biol.* 32(1957): 330–47; W. W. Howells, 'Explaining Modern Man: Evolutionists *versus* Migrationists', *J. Human Evol.* 5(1976): 477–95. Note the rather bizarre title of Howells' paper, distinguishing between evolution and migration as if they were diametrically opposed processes. The idea of migration, which we have already seen has only served to confuse the understanding of the process of geographical expansion, is evident from the title and has prevailed to the present.
5. R. L. Cann, M. Stoneking, and A. C. Wilson, 'Mitochondrial DNA and Human Evolution', *Nature* 325(1987): 31–6.
6. P. Mellars and C. Stringer (eds), *The Human Revolution: Behavioural and Biological Perspectives in the Origins of Modern Humans* (Edinburgh: Edinburgh University Press, 1989).
7. Mostly known now as the Out-of-Africa 2 model. The '2' is intended to differentiate the second expansion from Africa from the earlier one by *Homo erectus*. Lahr and Foley ('Multiple Dispersals and Modern Human Origins', *Evol. Anthropol.* 3(1994): 48–60) presented a more biologically enlightened view when they proposed the idea of multiple dispersals from Africa.
8. C. Stringer and R. McKie, *African Exodus: The Origins of Modern Humanity* (London: Jonathan Cape, 1996).
9. A. Hrdlička, 'The Neanderthal Phase of Man; The Huxley Memorial Lecture for 1927', *The Journal of the Royal Anthropological Institute of Great Britain and Ireland*, 57(1927): 249–74. Weindereich, ' "Neanderthal Man" ', and *idem, Apes, Giants and Men* (Chicago: University of Chicago Press, 1946); and later C. S. Coon, *The Origin of Races* (New York: Knopf, 1962) and *The Living Races of Man* (New York: Knopf, 1965), developed versions of the theory.

10. C. Loring Brace, 'The Fate of the "Classic" Neanderthals: A Consideration of Hominid Catastrophism', *Curr. Anthropol.* 5(1964): 3–43.

11. D. S. Brose and M. H. Wolpoff, 'Early Upper Palaeolithic Man and Late Middle Palaeolithic Tools', *Amer. Anthropol.* 73(1971): 1156–94; A. G. Thorne and M. H. Wolpoff, 'Regional Continuity in Australasian Pleistocene Hominid Evolution', *Am. J. Phys. Anthropol.* 55(1981): 337–49.

12. C. B. Stringer, 'Population Relationships of Later Pleistocene Hominids: A Multivariate Study of Available Crania', *J. Archaeol. Sci.* 1(1974): 317–42; C. B. Stringer and P. Andrews, 'Genetic and Fossil Evidence for the Origin of Modern Humans', *Science* 239(1974): 1263–8.

13. C. Finlayson and J. S. Carrión, 'Rapid Ecological Turnover and Its Impact on Neanderthal and Other Human Populations', *Trends Ecol. Evol.* 22(2007): 213–22.

14. The specimen is dated to between 33.5 and 35.5 thousand years ago. Calibrated dates, questionable for these early ages, place the specimen between 39 and 42 thousand years ago (see also n. 5 in Chapter 4); H. Shang et al., 'An Early Modern Human from Tianyuan Cave, Zhoukoudian, China', *Proc. Natl. Acad. Sci. USA* 104(2007): 6573–8.

15. See, for example, S. McBrearty and A. S. Brooks, 'The Revolution That Wasn't: A New Interpretation of the Origin of Modern Human Behaviour', *J. Hum. Evol.* 39(2000): 453–563.

16. P. Mellars, K. Boyle, O. Bar-Yosef, and C. Stringer (eds), *Rethinking the Human Revolution* (Cambridge: McDonald Institute Monographs, 2007).

17. T. H. van Andel, W. Davies, and B. Weninger, 'The Human Presence in Europe during the Last Glacial Period I: Human Migrations and the Changing Climate', in T. H. van Andel and W. Davies (eds), *Neanderthals and Modern Humans in the European Landscape during the Last Glaciation* (Cambridge: McDonald Institute Monographs, 2004), 31–56.

18. Dansgaard-Oeschger (DO) events were moments of rapid global warming of between 5 and 10 °C in a few decades followed by gradual cooling. Heinrich events (HE) were short-lived periods of extreme cold related to ice-rafting in the North Atlantic. HE depressed already cold temperatures by between 3 and 6 °C. Ten DO and three HE occurred between 50 and 30 thousand years ago; W. J. Burroughs, *Climate Change in Prehistory: The End of the Reign of Chaos* (Cambridge: Cambridge University Press, 2005).

19. N. W. Rutter et al., 'Correlation and Interpretation of Paleosols and Loess across European Russia and Asia over the Last Interglacial–Glacial Cycle', *Quat. Res.* 60(2003): 101–9.

20. J. Brigham-Grette et al., 'Chlorine-36 and 14C Chronology Support a Limited Last Glacial Maximum across Central Chukotka, Northeastern Siberia, and No Beringian Ice Sheet', *Quat. Res.* 59(2003): 386–98.

21. M. G. Grosswald, 'Late Weichselian Ice Sheets in Arctic and Pacific Siberia', *Quat. Int.* 45–6(1998): 3–18; M. G. Grosswald and T. J. Hughes, 'The Russian

Component of an Arctic Ice Sheet during the Last Glacial Maximum', *Quat. Sci. Rev.* 21(2002): 121–46.

22. A. N. Rudoy, 'Glacier-Dammed Lakes and Geological Work of Glacial Superfloods in the Late Pleistocene, Southern Siberia, Altai Mountains', *Quat. Int.* 87(2002): 119–40.

23. Finlayson and Carrión, 'Rapid Ecological Turnover'.

24. The last Neanderthals, ambush hunters par excellence, are recorded from Gibraltar between 28 and 24 kyr; C. Finlayson et al., 'Late Survival of Neanderthals at the Southernmost Extreme of Europe', *Nature* 443(2006): 850–3. They were not the only ambush predators to suffer the onslaught of the glacials. The leopard (*Panthera pardus*) survived in isolated southern pockets into historical times but its range was severely restricted between 50 and 30 kyr; E, R. S. Sommer and N. Benecke, 'Late Pleistocene and Holocene Development of the Felid Fauna (*Felidae*) of Europe: A Review', *J. Zool.* 269(2006): 7–19. The lion (*Panthera leo*) seems to have survived to the end of the last Ice Age in western Europe and its ability to hunt in open, treeless, habitats may have been an advantage; A. J. Stuart, 'Mammalian Extinctions in the Late Pleistocene of Northern Eurasia and North America', *Biol. Rev. Camb. Philos. Soc.* 66(1991): 453–562. The sabre-tooth cat (*Homotherium latidens*) was severely restricted in geographical range and is last recorded in western Europe 28 kyr; J. W. F. Reumer, 'Late Pleistocene Survival of the Sabre-Toothed Cat *Homotherium* in Northwestern Europe', *J. Vert. Paleontol.* 23(2003): 260–2. The Eurasian jaguar (*Panthera gombaszoegensis*), a species more dependent on dense woodland than the other species, just managed to survive to the end of the Middle Pleistocene; C. Guérin and M. Patou-Mathis, *Les grands mammifères plio-pleistocenes d'Europe* (Paris: Masson, 1997).

25. M. A. Cronin, S. C. Amstrup, and G. W. Garner, 'Interspecific and Intraspecific Mitochondrial DNA Variation in North American Bears (*Ursus*)', *Can. J. Zool.* 69(1991): 2985–92; S. L. Talbot and G. F. Shields, 'Phylogeography of Brown Bears (*Ursus arctos*) of Alaska and Paraphyly within the Ursidae', *Mol. Phylog. Evol.* 5(1996): 477–94.

26. Finlayson and Carrión, 'Rapid Ecological Turnover'.

27. Van Andel, Davies, and Weninger, 'The Human Presence in Europe'.

28. Finlayson et al., 'Late Survival of Neanderthals'.

29. M. P. Richards et al., 'Stable Isotope Evidence for Increasing Dietary Breadth in the European mid-Upper Paleolithic', *Proc. Natl. Acad. Sci. USA* 98(2001): 6528–32; E. Trinkaus et al., 'An Early Modern Human from the Peştera cu Oase, Romania', *Proc. Natl. Acad. Sci. USA* 100(2003): 11231–6; E. M. Wild et al., 'Direct Dating of Early Upper Palaeolithic Human Remains from Mladeč, *Nature* 435(2005): 332–5; A. Soficaru, A. Dobos, and E. Trinkaus, 'Early Modern Humans from the Peştera Muierii, Baia de Fier, Romania', *Proc. Natl. Acad. Sci. USA* 103(2006): 17196–201.

30. E. Trinkaus, 'Early Modern Humans', *Ann. Rev. Anthropol.* 34(2005): 207–30.

31. P. Underhill et al., 'The Phylogeography of Y Chromosome Binary Haplotypes and the Origins of Modern Human Populations', *Ann. Hum. Genet.* 65(2001): 43–62; P. Forster, 'Ice Ages and the Mitochondrial DNA Chronology of Human Dispersals: A Review', *Phil. Trans. Roy. Soc. Lond. B.* 359(2004): 255–64.

32. P. Mellars, 'Neanderthals and the Modern Human Colonization of Europe', *Nature* 432(2004): 461–5.

33. A number of cultures with technologies that appear to traverse the Middle–Upper Palaeolithic boundary, combining elements of both, are described in the archaeological literature as transitional cultures, technologies, or industries. J. Zilhao and F. d'Errico, 'La nouvelle "bataille aurignacienne": Une révision critique de la chronologie du Châtelperronien et de l'Aurignacien ancien', *L'Anthropologie* 104(2000): 17–50; J. Zilhao et al., 'Analysis of Aurignacian Interstratification at the Châtelperronian-Type Site and Implications for the Behavioral Modernity of Neandertals', *Proc. Natl. Acad. Sci. USA* 103(2006): 12643–8.

34. The Châtelperronian has been associated with Neanderthal remains in the French sites of St.-Césaire and the Grotte du Renne. Some authors question the association between the human remains and the material culture; O. Bar-Yosef, 'Defining the Aurignacian', in O. Bar-Yosef and J. Zilhao (eds), *Towards a Definition of the Aurignacian*, Trabalhos de Arqueologia, 46 (Portugal: IPA, 2006), 11–18.

35. Finlayson and Carrión, 'Rapid Ecological Turnover'.

36. Mellars, 'Neanderthals and the Modern Human Colonization'.

37. N. J. Conard, P. M. Grootes, and F. H. Smith, 'Unexpectedly Recent Dates for Human Remains from Vogelherd', *Nature* 430(2004): 198–201.

38. Finlayson and Carrión, 'Rapid Ecological Turnover'.

39. T. Goebel, A. Derevianko, and V. T. Petrin, 'Dating the Middle-to-Upper Paleolithic Transition at Kara-Bom', *Curr. Anthropol.* 34(1993): 452–8; M. Otte and A. Derevianko, 'Transformations Techniques au Paléolithique de l'Altaï', *Anthropol. Prehist.* 107(1996): 131–43; Y. V. Kuzmin and L. A. Orlova, 'Radiocarbon Chronology of the Siberian Paleolithic', *J. World Prehist.* 12(1998): 1–53; J. K. Kozlowski, 'The Problem of Cultural Continuity between the Middle and the Upper Paleolithic in Central and Eastern Europe', in O. Bar-Yosef and D. Pilbeam (eds), *The Geography of Neandertals and Modern Humans in Europe and the Greater Mediterranean*, Peabody Museum Bulletin 8 (Cambridge, MA: Harvard University Press, 2000), 77–105; P. Pavlov, J. I. Svendsen, and S. Indrelid, 'Human Presence in the European Arctic Nearly 40,000 years ago', *Nature* 413(2001): 64–7; P. Pavlov, W. Roebroeks, and J. I. Svendsen, 'The Pleistocene Colonization of Northeastern Europe: A Report on Recent Research, *J. Hum. Evol.* 47(2004): 3–17; J. F. Hoffecker, 'Innovation and Technological Knowledge in the Upper Paleolithic of Northern Eurasia', *Evol. Anthropol.* 14(2005): 186–98; M. V. Anikovich et al., 'Early Upper Paleolithic in Eastern Europe and Implications for the Dispersal of Modern Humans', *Science* 315(2007): 223–6.

40. Finlayson and Carrión, 'Rapid Ecological Turnover'.
41. B. Blades, 'Aurignacian Settlement Patterns in the Vézère Valley', *Curr. Anthropol.* 40(1999): 712–18.
42. J. R. M. Allen et al., 'Rapid Environmental Changes in Southern Europe during the Last Glacial Period', *Nature* 400(1999): 740–3.
43. R. Musil, 'The Middle and Upper Palaeolithic Game Suite in Central and Southeastern Europe', in van Andel and Davies (eds), *Neanderthals and Modern Humans*, 167–90.
44. J. Chlachula, 'Pleistocene Climate Change, Natural Environments and Palaeolithic Occupation of the Angara–Baikal Area, East Central Siberia', *Quat. Int.* 80–1(2001): 69–92; J. Chlachula, 'Pleistocene Climate Change, Natural Environments and Palaeolithic Occupation of the Upper Yenisei Area, South-Central Siberia', *Quat. Int.* 80–1(2001): 101–30; J. Chlachula, 'Pleistocene Climate Change, Natural Environments and Palaeolithic Occupation of the Altai Area, West-Central Siberia', *Quat. Int.* 80–1(2001): 131–67.
45. The fauna of the Bykovsky Peninsula was made up of woolly mammoth, woolly rhino (rare), horse, reindeer, steppe bison and musk ox; L. Schirrmeister et al., 'Paleoenvironmental and Paleoclimatic Records from Permafrost Deposits in the Arctic Region of Northern Siberia', *Quat. Int.* 89(2002): 97–118.
46. Pavlov et al., 'Human Presence in the European Arctic'; V. V. Pitulko et al., 'The Yana RHS Site: Humans in the Arctic before the Last Glacial Maximum', *Science* 303(2004): 52–6.
47. The Mediterranean Sea marked its western limit.
48. R. Rabinovich, 'The Levantine Upper Palaeolithic Faunal Record', in A. N. Goring-Morris and A. Belfer-Cohen (eds), *More than Meets the Eye: Studies on Upper Palaeolithic Diversity in the Near East* (Oxford: Oxbow Books, 2003), 33–48.
49. O. Bar-Yosef, 'The Middle and Early Upper Paleolithic in Southwest Asia and Neighboring Regions', in Bar-Yosef and Pilbeam (eds), *Geography of Neanderthals*, 107–56.
50. Finlayson and Carrión, 'Rapid Ecological Turnover'.
51. S. Oppenheimer, *Out of Eden: The Peopling of the World* (London: Robinson, 2004).
52. Finlayson and Carrión, 'Rapid Ecological Turnover'.
53. The Aterian culture that spread from Arabia to Morocco, Chapter 5.
54. J-J Hublin et al., 'A Late Neanderthal Associated with Upper Palaeolithic Artefacts', *Nature* 381(1996): 224–6.
55. F. d'Errico et al., 'Neanderthal Acculturation in Western Europe? A Critical Review of the Evidence and Its Interpretation', *Curr. Anthropol.* 39(1998): S1-S44.
56. F. d'Errico, 'The Invisible Frontier. A Multiple Species Model for the Origin of Behavioral Modernity', *Evol. Anthropol.* 12(2003): 186–202; J. Zilhão, 'The

Emergence of Ornaments and Art: An Archaeological Perspective on the Origins of "Behavioral Modernity", *J. Archaeol. Res.* 15(2007): 1–54.

57. The Neanderthal skull from Forbes' Quarry in Gibraltar was actually found eight years before the Neander Valley specimen in Germany but it was not formally given a scientific name.

58. P. Mellars, 'The Neanderthal Problem Continued', *Curr. Anthropol.* 40(1999): 341–64; J. Zilhão and F. d'Errico, 'The Chronology and Taphonomy of the Earliest Aurignacian and Its Implications for the Understanding of Neandertal Extinction', *J. World Prehist.* 13(1999): 1–68; Zilhão and d'Errico, 'La nouvelle "bataille aurignacienne" '; F. d'Errico et al., 'Many Awls in Our Argument: Bone Tool Manufacture and Use in the Châtelperronian and Aurignacian Levels of the Grotte du Renne at Arcy-sur-Cure', in J. Zilhão and F. d'Errico (eds), *The Chronology of the Aurignacian and of the Transitional Technocomplexes*, Trabalhos de Arqueologia 33 (Portugal: IPA, 2003), 247–70; J. Zilhão and F. d'Errico, 'The Chronology of the Aurignacian and Transitional Technocomplexes: Where Do We Stand?', ibid. 313–49; J. Zilhão and F. d'Errico, 'An Aurignacian "Garden of Eden" in Southern Germany? An Alternative Interpretation of the Giessenklösterle and Critique of the *Kulturpumpe* Model', *Paleo* 15(2003): 69–86; B. Gravina, P. Mellars, and C. Bronk Ramsey, 'Radiocarbon Dating of Interstratified Neanderthal and Early Modern Human Occupations at the Châtelperronian Type-Site', *Nature* 438(2005): 51–6; P. Mellars, 'The Impossible Coincidence: A Single-Species Model for the Origins of Modern Human Behavior in Europe', *Evol. Anthropol.* 14(2005): 12–27; P. Mellars, 'Archeology and the Dispersal of Modern Humans in Europe: Deconstructing the "Aurignacian" ', *Evol. Anthropol.* 15(2006): 167–82; J. Zilhão, 'Aurignacian, Behavior, Modern: Issues of Definition in the Emergence of the European Upper Paleolithic', in Bar-Yosef and Zilhão (eds), *Towards a Definition of the Aurignacian*, 53–69; J. Zilhão et al., 'Analysis of Aurignacian Interstratification at the Châtelperronian-Type Site and Implications for the Behavioral Modernity of Neandertals', *Proc. Natl. Acad. Sci. USA* 103(2006): 12643–8; P. Mellars, B. Gravina, and C. Bronk Ramsey, 'Confirmation of Neanderthal/Modern Human Interstratification at the Châtelperronian Type-Site', *Proc. Natl. Acad. Sci. USA* 104(2007): 3657–62.

59. C. Finlayson, *Neanderthals and Modern Humans: An Ecological and Evolutionary Perspective* (Cambridge: Cambridge University Press, 2004).

60. The authors of this paper and subsequent ones talk of genetic admixture rather than hybrids, presumably because from their standpoint Neanderthals and Ancestors were the same species. A hybrid would be the product of the two species not one. I will not split hairs and refer to hybrids as the products of Neanderthal–Ancestor mating without judging whether we are dealing with one or two species. The key references to the Lagar Velho hybrid are C. Duarte et al., 'The Early Upper Palaeolithic Human Skeleton from the Abrigo do Lagar Velho (Portugal) and Modern Human Emergence in Iberia', *Proc. Natl.*

Acad. Sci. USA 96(1999): 7604–09; J. Zilhão and E. Trinkaus (eds), *Portrait of the Artist as a Child: The Gravettian Human Skeleton from the Abrigo do Lagar Velho and Its Archeological Context*, Trabalhos de Arqueologia 22 (Portugal: IPA, 2002).

61. I. Tattersall and J. Schwartz, 'Hominids and Hybrids: The Place of Neanderthals in Human Evolution', *Proc. Natl. Acad. Sci. USA* 96(1999): 7117–19.

62. Finlayson et al., 'Late Survival of Neanderthals'.

63. Soficaru, Dobos, and Trinkaus, 'Early Modern Humans'.

64. E. Trinkaus, 'European Early Modern Humans and the Fate of the Neandertals', *Proc. Natl. Acad. Sci. USA* 104(2007): 7367–72.

65. R. R. Ackermann, J. Rogers, and J. M. Cheverud, 'Identifying the Morphological Signatures of Hybridization in Primate and Human Evolution', *J. Hum. Evol.* 51(2006): 632–45.

66. M. Krings et al., 'Neandertal DNA Sequences and the Origin of Modern Humans', *Cell* 90(1997): 19–30; M. Krings et al., 'DNA Sequence of the mitochondrial Hypervariable Region II from the Neandertal Type Specimen', *Proc. Natl. Acad. Sci. USA* 96(1999): 5581–5; I. V. Ovchinnikov et al., 'Molecular Analysis of Neanderthal DNA from the Northern Caucasus', *Nature* 404(2000): 490–3; D. Caramelli et al., 'Evidence for a Genetic Discontinuity between Neandertals and 24,000-Year-Old Anatomically Modern Europeans', *Proc. Natl. Acad. Sci. USA* 100(2003): 6593–7; C. Lalueza-Fox et al., 'Neandertal Evolutionary Genetics; Mitochondrial DNA Data from the Iberian Peninsula', *Mol. Biol. Evol.* 22(2005): 1077–81; R. E. Green et al., 'Analysis of One Million Base Pairs of Neanderthal DNA', *Nature* 444(2006): 330–6; J. P. Noonan et al., 'Sequencing and Analysis of Neanderthal Genomic DNA', *Science* 314(2006): 1113–18.

67. D. Serre et al., 'No Evidence of Neandertal mtDNA Contribution to Early Modern Humans', *PLoS Biol.* 2(2004): e57; M. Currat and L. Excoffier, 'Modern Humans Did Not Admix with Neanderthals during Their Range Expansion into Europe', *PLoS Biol.* 2(2004): e421.

68. M. Ponce de León and C. Zollikofer, 'Neanderthal Cranial Ontogeny and Its Implications for Late Hominid Diversity', *Nature* 412(2001): 534–8.

Chapter 7—Africa in Europe—A Mediterranean Serengeti

1. C. Finlayson, *Al-Andalus: How Nature Has Shaped History* (Málaga: Santana Books, 2007).

2. C. Finlayson et al., 'Late Survival of Neanderthals at the Southernmost Extreme of Europe', *Nature* 443(2006): 850–3.

3. G. Finlayson et al. 'Caves as Archives of Ecological and Climatic Changes in the Pleistocene—The Case of Gorham's Cave, Gibraltar', *Quat. Int.* 181(2008): 55–63.

4. G. Finlayson, 'Climate, Vegetation and Biodiversity—A Multiscale Study of the South of the Iberian Peninsula', PhD thesis, University of Anglia Ruskin, Cambridge, 2006.

5. My good friend and colleague Doug Larson from the University of Guelph in Canada first exclaimed 'My goodness, this was Neanderthal City' when he first saw the line of Gibraltar caves.

6. The radiocarbon dates leave a period of 5.5 thousand years between the last Neanderthals and the first ancestors. Calibrating these dates from radiocarbon to calendrical years would put the last Neanderthals at 28–29 thousand years ago and the first ancestors at 21–22 thousand years ago, making the period when the cave was unoccupied anything between 6 and 8 thousand years. The historical levels at Gorham's start in the 8th century BC (Phoenician) and end in the 14th century AD (Muslim).

7. The climate today has mean annual temperatures between 17 and 19 °C and annual rainfall between 600 and 1000 mm. For the entire last glacial cycle annual temperatures ranged between 13 and 19 °C and annual rainfall between 350 and 1000 mm. Finlayson, 'Climate, Vegetation and Biodiversity'.

8. Ibid.

9. 1 hectare = 100 × 100 metres. Habitat structure is the three-dimensional arrangement of objects in space and the measurements included tree, shrub, and grass cover and height, tree density, and so on. Put together, a numerical description of the habitat was possible.

10. In Doñana, corrales are patches of stone pine woodland surrounded by the moving sand dunes. They are eventually covered up completely and die. New pines set seed and grow where the dunes are inactive and new woods form. They will, in turn, be engulfed by the sands when a change of wind gets them on the move once more. The stone pine–sand dune system of Doñana is a highly dynamic one.

11. R. G. Klein, *The Human Career: Human Biological and Cultural Origins* (Chicago: Chicago University Press, 1999).

12. M. C. Stiner et al., 'Paleolithic Population Growth Pulses Evidenced by Small Animal Exploitation', *Science* 283(1999): 190–4; M. C. Stiner, N. D. Munro, and T. A. Surovell, 'The Tortoise and the Hare: Small-Game Use, the Broad Spectrum Revolution, and Paleolithic Demography', *Curr. Anthropol.* 41(2000): 39–74.

13. Apart from the inherent problems in demonstrating such an assertion, the statistics published did not actually support the claim so that a decrease in shell size through time could not be conclusively demonstrated. The problems were compounded because sites from different geographical areas and time periods were compared and this made it impossible to say whether any changes had to do with time or simply because different regions were being compared. To make matters worse limpets of different species, known to be of different size, were all lumped as 'limpets'. Let us imagine what would happen if we got eagles and sparrows from different time periods. One time period had many eagles and few sparrows so our average measurement of birds would show them to be large; the next period was dominated by sparrows so the average

measurement would be low. Would we assume that birds had got smaller or, instead, that we were measuring different things?

14. Anyone with any knowledge of land tortoise natural history knows that tortoises may move slowly but they are hard to find. They hide in dense vegetation and they spend the winter months hibernating—hardly a recipe for 'easy' prey.

15. C. Finlayson, *Birds of the Strait of Gibraltar* (London: Academic Press, 1992).

16. Kimberly Brown, a PhD student at Cambridge University, has shown that Neanderthals brought many different kinds of birds back to the cave to be eaten. They included partridges, quails, and ducks.

17. Charred seeds have been found inside hearths that had been made by the Neanderthals.

18. C. B. Stringer et al., 'Neanderthal Exploitation of Marine Mammals in Gibraltar', *Proc. Natl. Acad. Sci. USA* 105(2008): 14319–24.

19. In recent years the ratio of carbon and nitrogen isotopes in human teeth and bone has been used to reconstruct the diet of prehistoric people. Carbon (^{12}C and ^{13}C) and nitrogen (^{14}N and ^{15}N) have two stable isotopes each and small differences in the ratios of these isotopes have been used to identify the diet of particular individuals. H. Bocherens et al., 'Isotopic Biogeochemistry (^{13}C, ^{15}N) of Fossil Vertebrate Collagen: Application to the Study of a Past Food Web Including Neandertal Man', *J. Hum. Evol.* 20(1991): 481–92; M. Fizet et al., 'Effect of Diet, Physiology and Climate on Carbon and Nitrogen Stable Isotopes of Collagen in a Late Pleistocene Anthropic Palaecosystem: Marillac, Charente, France', *J. Archaeol. Sci.* 22(1995): 67–79; H. Bocherens et al., 'Palaeoenvironmental and Palaeodietary Implications of Isotopic Biogeochemistry of Last Interglacial Neanderthal and Mammal Bones from Scladina Cave (Belgium)', *J. Archaeol. Sci.* 26(1999): 599–607; M. Richards et al., 'Neanderthal Diet at Vindija and Neanderthal Predation: The Evidence from Stable Isotopes', *Proc. Natl. Acad. Sci. USA* 97(2000): 7663–6; M. Richards et al., 'Stable Isotope Evidence for Increasing Dietary Breadth in the European Mid-Upper Paleolithic', *Proc. Natl. Acad. Sci. USA* 98(2001): 6528–32; D. Drucker and H. Bocherens, 'Carbon and Nitrogen Stable Isotopes as Tracers of Change in Diet Breadth during Middle and Upper Palaeolithic in Europe', *Int. J. Osteoarch.* 14(2004): 162–77; H. Bocherens et al., 'Isotopic Evidence for Diet and Subsistence Pattern of the Saint-Césaire I Neanderthal: Review and Use of a Multisource Mixing Model', *J. Hum. Evol.* 49(2005): 71–87.

20. R. Jennings, 'Neanderthal and Modern Human Occupation Patterns in Southern Iberia during the Late Pleistocene Period', DPhil Thesis, University of Oxford, 2006.

21. N. García and J. L. Arsuaga, 'Late Pleistocene Cold-Resistant Faunal Complex: Iberian Occurrences, in M. Blanca Ruiz Zapata et al. (eds), *Quaternary Climatic Changes and Environmental Crises in the Mediterranean Region* (Madrid: Universidad de Alcala de Henares, 2003), 149–59.

22. M. Vaquero et al., 'The Neandertal–Modern Human Meeting in Iberia: A Critical Review of the Cultural, Geographical and Chronological Data', in N. J. Conard (ed.), *When Neanderthals and Modern Humans Met* (Tübingen: Kerns Verlag, 2006), 419–39.

23. The mountain ranges of south-eastern Spain are the highest of the Iberian Peninsula, surpassing 3000 metres in the Sierra Nevada.

24. F. J. Jiménez-Espejo et al., 'Climate Forcing and Neanderthal Extinction in Southern Iberia: Insights from a Multiproxy Marine Record', *Quat. Sci. Rev.* 26(2007): 836–52.

25. The climatic cause of the demise of the last Neanderthals was subsequently taken, erroneously, to mean the cause of the Neanderthal extinction. A paper tried to compare the climate signals in the region with the last reported dates from Gorham's Cave and concluded that climate was not unduly harsh. The mistake was that the authors failed to recognize that the published dates were late survival and not disappearance dates. That the dates coincided with benign conditions is precisely what would have been expected. P. C. Tzedakis et al., 'Placing Late Neanderthals in a Climatic Context', *Nature* 449(2007): 206–8.

Chapter 8—One Small Step for Man ...

1. E. Trinkaus, 'Early Modern Humans', *Ann. Rev. Anthropol.* 34(2005): 207–30.

2. J. T. Kerr and L. Packer, 'Habitat Heterogeneity as a Determinant of Mammal Species Richness in High-Energy Regions', *Nature* 385(1997): 252–4; C. Finlayson, *Neanderthals and Modern Humans: An Ecological and Evolutionary Perspective* (Cambridge: Cambridge University Press, 2004).

3. C. Finlayson and J. S. Carrión, 'Rapid Ecological Turnover and Its Impact on Neanderthal and Other Human Populations', *Trends Ecol. Evol.* 22(2007): 213–22.

4. S. Wells et al., 'The Eurasian Heartland: A Continental Perspective on Y-Chromosome Diversity', *Proc. Natl. Acad. Sci. USA* 98(2001): 10244–9.

5. M. B. Richards et al., 'Phylogeography of Mitochondrial DNA in Western Europe', *Ann. Hum. Genet.* 62(1998): 241–60; P. A. Underhill et al., 'The Phylogeography of Y Chromosome Binary Haplotypes and the Origins of Modern Human Populations', *Ann. Hum. Genet.* 65(2001): 43–62; P. Forster, 'Ice Ages and the Mitochondrial DNA Chronology of Human Dispersals: A Review', *Phil. Trans. Roy. Soc. Lond. B.* 359(2004): 255–64.

6. M. Anikovich, 'Early Upper Paleolithic Industries of Eastern Europe', *J. World Prehist.* 6(1992): 205–45; T. Goebel et al., 'Dating the Middle-to-Upper Paleolithic Transition at Kara-Bom', *Curr. Anthropol.* 34(1993): 452–8; T. Goebel and M. Aksenov, 'Accelerator Radiocarbon Dating of the Initial Upper Palaeolithic in Southeast Siberia', *Antiquity* 69(1995): 349–57; M. Otte and A. Derevianko, 'Transformations Techniques au Paléolithique de l'Altaï (Sibérie)', *Anthropol. et Préhist.* 107(1996): 131–43; Y. V. Kuzmin, 'The Colonization of Eastern

Siberia: an Evaluation of the Paleolithic Age Radiocarbon Dates', *J. Archaeol. Sci.* 23(1996): 577–85; P. J. Brantingham et al., 'The Initial Upper Paleolithic in Northeast Asia', *Curr. Anthropol.* 42(2001): 735–47; P. Pavlov, J. I. Svendsen, and S. Indrelid, 'Human Presence in the European Arctic Nearly 40,000 years ago', *Nature* 413(2001): 64–7; P. Pavlov, W. Roebroeks, and J. I. Svendsen, 'The Pleistocene Colonization of Northeastern Europe: A Report on Recent Research, *J. Hum. Evol.* 47(2004): 3–17; M. V. Anikovich et al., 'Early Upper Paleolithic in Eastern Europe and Implications for the Dispersal of Modern Humans', *Science* 315(2007): 223–6.

7. Anikovich, 'Early Upper Paleolithic Industries'; Otte and Derevianko, 'Transformations Techniques'; V. Y. Cohen and V. N. Stepanchuk, 'Late Middle and Early Upper Paleolithic Evidence from the East European Plain and Caucasus: A New Look at Variability, Interactions, and Transitions', *J. World Prehist.* 13(1999): 265–319; Brantingham et al., 'The Initial Upper Paleolithic in Northeast Asia'; V. P. Chabai, 'The Chronological and Industrial Variability of the Middle to Upper Paleolithic Transition in Eastern Europe', in J. Zilhao and F. d'Errico (eds), *The Chronology of the Aurignacian and of the Transitional Technocomplexes. Dating, Stratigraphies, Cultural Implications*, Trabalhos de Arqueologia 33 (Portugal: IPA, 2003), 71–86; Anikovich et al., 'Early Upper Paleolithic in Eastern Europe'.

8. Wells et al., 'The Eurasian Heartland'; S. Wells, *The Journey of Man: A Genetic Odyssey* (London: Penguin, 2002).

9. O. Semino et al., 'The Genetic Legacy of Paleolithic *Homo sapiens sapiens* in Extant Europeans: A Y Chromosome Perspective', *Science* 290(2001): 1155–9; Wells et al., 'The Eurasian Heartland'; Wells, *The Journey of Man*.

10. J. Diamond, *Guns, Germs and Steel. A Short History of Everybody for the Last 13,000 Years* (London: Jonathan Cape, 1997).

11. S. McBrearty and A. S. Brooks, 'The Revolution That Wasn't: A New Interpretation of the Origin of Modern Human Behaviour', *J. Hum. Evol.* 39(2000): 453–563.

12. Finlayson, *Neanderthals and Modern Humans*.

13. Bone and antler tools are also found in Aurignacian (unknown makers) and Châtelperronian (made by Neanderthals) communities that were in contact with, though less committed to, the plains than the Gravettians. Such tools occasionally appeared earlier in the African archaeological record; McBrearty and Brooks, 'The Revolution That Wasn't'.

14. C. Gamble, *The Palaeolithic Settlement of Europe* (Cambridge: Cambridge University Press, 1986).

15. C. Gamble, *The Palaeolithic Societies of Europe* (Cambridge: Cambridge University Press, 1999)

16. E. Carbonell and I. Roura, *Abric Romaní Nivell I. Models d'ocupació de curta durada de fa 46.000 anys a la Cinglera del Capelló* (Capellades, Anoia, Barcelona: Universitat Rovira I Virgili, Tarragona, 2002).

17. J. Svoboda, S. Péan, and P. Wojtal, 'Mammoth Bone Deposits and Subsistence Practices during Mid-Upper Palaeolithic in Central Europe: Three Cases from Moravia and Poland', *Quat. Int.* 126–8(2005): 209–21.

18. O. Soffer, 'Storage, Sedentism and the Eurasian Palaeolithic Record', *Antiquity* 63(1989): 719–32; O. Soffer et al., 'Cultural Stratigraphy at Mezhirich, an Upper Palaeolithic Site in Ukraine with Multiple Occupations', *Antiquity* 71(1997): 48–62.

19. P. Ward and A. Zahavi, 'The Importance of Certain Assemblages of Birds as "Information Centers" for Food Finding', *Ibis* 115(1973): 517–34.

20. C. Marean et al., 'Early Human Use of Marine Resources and Pigment in South Africa during the Middle Pleistocene', *Nature* 449(2007): 905–9.

21. Paintings in Chauvet Cave, France, indicate two periods of painting: 32–30 and 27–26 thousand years ago. The first period has been linked to the Aurignacians (presumed, without conclusive evidence, to be Ancestors) purely on the basis of the dates but there are Gravettian sites in France that go back to 30–29 thousand years ago. Many of the radiocarbon dates have been taken from bone. Recent work has shown that bone samples, when pretreated by a new system of ultra-filtration, give dates that may be between 2 and 7 thousand years older than the initial estimates. This means that many French Gravettian sites would be contemporary with, or even predate, Chauvet Cave's art. J. Clottes, *Chauvet Cave. The Art of Earliest Times* (Salt Lake City: University of Utah Press, 2003); P. Mellars, 'A New Radiocarbon Revolution and the Dispersal of Modern Humans in Eurasia), *Nature* 439(2006): 931–5.

22. In Dolni Věstonice, Czech Republic, the ceramic inventory consists of over 5,000 artefacts that were fired at temperatures between 500 and 800 °C between 28 and 24 thousand years ago. The prime source of raw material for the ceramics was the loess, a fine wind-blown sediment that covered huge areas of northern Eurasia in the Pleistocene; P. B. Vandiver et al., 'The Origins of Ceramic Technology at Dolni Věstonice, Czechoslovakia', *Science* 246(1989): 1002–8.

23. Y. V. Kuzmin, 'The Earliest Centres of Pottery Origin in the Russian Far East and Siberia: Review of Chronology for the Oldest Neolithic Cultures', *Documenta Praehistorica* 29(2002): 37–46.

24. F. d'Errico, 'The Invisible Frontier. A Multiple Species Model for the Origin of Behavioral Modernity', *Evol. Anthropol.* 12(2003): 186–202.

25. J. V. Turcios, *Maestros subterraneos: Las tecnicas del arte Paleolitico* (Madrid: Celeste, 1995).

26. O. Soffer, 'Artistic Apogees and Biological Nadirs: Upper Paleolithic Cultural Complexity Reconsidered', in M. Otte (ed.), *Nature et Culture* (Liège: ERAUL, 1995), 615–27.

27. Finlayson and Carrión, 'Rapid Ecological Turnover'.

28. H. H. Draper, 'The Aboriginal Eskimo Diet in Modern Perspective', *Amer. Anthropol.* 79(1977): 309–16.

29. J. M. Adovasio et al., 'Perishable Industries from Dolní Věstonice I: New Insights into the Nature and Origin of the Gravettian', *Archaeol., Ethnol., Anthropol., Eurasia* 2(2001): 48–64.

30. Palaeoanthropologist Yoel Rak at the Hebrew University in Jerusalem links differences between Neanderthal and Ancestor pelvises to locomotion, the pelvis of Ancestors being better able at cushioning stresses of walking for long distances.

Chapter 9—Forever Opportunists

1. A rough calculation converts the distance of 4,500 kilometres in a millennium, using the calculation of a human generation time of 20 years that we used in the Prologue, to a 90 kilometres/generation rate of spread which is much faster than the 60 kilometres/generation that we calculated for the Africa-Australia spread. The actual rate of spread may have been faster given that the dates for Gravettian appearance in different regions have large errors. Although very approximate the difference is large enough to suggest that the Eurasian Plains people spread much faster than their predecessors north of the Indian Ocean.

2. C. Finlayson, *Neanderthals and Modern Humans: An Ecological and Evolutionary Perspective* (Cambridge: Cambridge University Press, 2004).

3. E. Trinkaus, 'The Neanderthals and Modern Human Origins', *Ann. Rev. Anthropol.* 15(1986): 193–218; T. M. Smith et al., 'Rapid Dental Development in a Middle Paleolithic Belgian Neanderthal', *Proc. Natl. Acad. Sci. USA* 104(2007): 20220–5.

4. Humans perform remarkably well at endurance running when compared to many animals, having a number of anatomical features suited for the purpose. Endurance running seems to be a feature of the genus *Homo* and may date back to 2 million years ago; D. M. Bramble and D. E. Lieberman, 'Endurance Running and the Evolution of *Homo*', *Nature* 432(2004): 345–52.

5. J. Clutton-Brock, *A Natural History of Domesticated Mammals* (London: Natural History Museum, 1999).

6. M. V. Sablin and G. A. Khlopachev, 'The Earliest Ice Age Dogs: Evidence from Eliseevichi I', *Curr. Anthropol.* 43(2002): 795–9.

7. C. Vilà et al., 'Multiple and Ancient Origins of the Domestic Dog', *Science* 276(1997): 1687–9.

8. Other than other humans, in cooperation.

9. C. Gamble, *The Palaeolithic Societies of Europe* (Cambridge: Cambridge University Press, 1999).

10. P. Clarke, *Where the Ancestors Walked* (Crow's Nest, NSW: Allen and Unwin, 2003).

11. V. V. Pitulko et al., 'The Yana RHS Site: Humans in the Arctic before the Last Glacial Maximum', *Science* 303(2004): 52–6.

12. S. Wells, *The Journey of Man: A Genetic Odyssey* (London: Penguin, 2002); S. Oppenheimer, *Out of Eden: The Peopling of the World* (London: Robinson, 2004); Y. V. Kuzmin and S. G. Keates, 'Dates Are Not Just Data: Paleolithic Settlement Patterns in Siberia Derived from Radiocarbon Records', *Amer. Antiquity* 70(2005): 773–89; T. D. Goebel, M. R. Waters, and H. O'Rourke, 'The Late Pleistocene Dispersal of Modern Humans in the Americas', *Science* 319(2008): 1497–502.

13. H. Shang et al., 'An Early Modern Human from Tianyuan Cave, Zhoukoudian, China', *Proc. Natl. Acad. Sci. USA* 104(2007): 6573–8.

14. That is, not showing any remaining archaic features. The term modern is somewhat misleading but is used in this book, sparingly, to avoid confusion with the bulk of the public literature. The reality is that all contemporary groups across the world at any time would have been, by definition, equally modern.

15. Goebel et al., 'The Late Pleistocene Dispersal'; A. Kitchen, M. M. Miyamoto, and C. J. Mulligan, 'A Three-Stage Colonization Model for the Peopling of the Americas', *PLoS ONE* 3(2008): e1596.

16. Ibid.

17. Radiocarbon dates are in calendar years, as are others referred to in this chapter, given that they are within the range of reliable calibration. T. D. Dillehay et al., 'Monte Verde: Seaweed, Food, Medicine, and the Peopling of South America', *Science* 320(2008): 784–6.

18. A. L. Martinez, '9,700 Years of Maritime Subsistence on the Pacific: An Analysis by Means of Bioindicators in the North of Chile', *Amer. Antiquity* 44(1979): 309–24; D. H. Sandweiss et al., 'Quebrada Jaguay: Early South American Maritime Adaptations', *Science* 281(1998): 1830–2; D. K. Keefer et al., 'Early Maritime Economy and El Niño Events at Quebrada Tacahuay, Peru', *Science* 281(1998): 1833–5; D. Jackson et al., 'Initial Occupation of the Pacific Coast of Chile during Late Pleistocene Times', *Curr. Anthropol.* 48(2007): 725–31.

19. Goebel et al., 'The Late Pleistocene Dispersal'.

20. D. J. Joyce, 'Chronology and New Research on the Schaefer Mammoth (?*Mammuthus primigenius*) Site, Kenosha County, Wisconsin, USA', *Quat. Int.* 142–3(2006): 44–57; Goebel et al., 'The Late Pleistocene Dispersal'. Evidence of such behaviour before 15 thousand years ago exists but is less secure.

21. W. J. Burroughs, *Climate Change in Prehistory: The End of the Reign of Chaos* (Cambridge: Cambridge University Press, 2005).

22. C. Finlayson and J. S. Carrión, 'Rapid Ecological Turnover and Its Impact on Neanderthal and Other Human Populations', *Trends Ecol. Evol.* 22(2007): 213–22.

23. Finlayson, *Neanderthals and Modern Humans*.

24. T. Pakenham, *The Scramble for Africa* (London: Abacus, 1992); H. Reynolds, *Why Weren't We Told? A Personal Search for the Truth about Our History* (Victoria: Penguin, 1999).

25. J. Diamond, *Guns, Germs and Steel. A Short History of Everybody for the Last 13,000 Years* (London: Jonathan Cape, 1997).

Chapter 10—The Pawn Turned Player

1. P. A. Underhill et al., 'The Phylogeography of Y Chromosome Binary Haplotypes and the Origins of Modern Human Populations', *Ann. Hum. Genet.* 65(2001): 43–62.

2. A. N. Goring-Morris and A. Belfer-Cohen (eds), *More Than Meets the Eye: Studies on Upper Palaeolithic Diversity in the Near East* (Oxford: Oxbow Books, 2003).

3. A. Belfer-Cohen and N. Goring-Morris, 'Why Microliths? Microlithization in the Levant', *Archaeol. Papers Amer. Anthropol. Assocn.* 12(2002): 57–68.

4. S. L. Kuhn, 'Pioneers of Microlithization: The "Proto-Aurignacian" of Southern Europe', *Archaeol. Papers Amer. Anthropol. Assocn.* 12(2002): 83–93. Such early attempts at producing microlithic technologies also appear in other regions, for example around 36 thousand years ago in Sri Lanka: K. A. R. Kennedy, *God-Apes and Fossil Men: Paleoanthropology of South Asia* (Ann Arbor: University of Michigan Press, 2000).

5. S. L. Kuhn and R. G. Elston, 'Thinking Small Globally', *Archaeol. Papers Amer. Anthropol. Assocn.* 12(2002): 1–7.

6. S. Mithen, *After the Ice: A Global Human History 20,000–5000 BC* (London: Weidenfeld and Nicolson, 2003).

7. D. Nadel and E. Werker, 'The Oldest Ever Brush Hut Plant Remains from Ohalo II, Jordan Valley, Israel (19,000 BP)', *Antiquity* 73(1999): 755–64; D. Nadel et al., 'Stone Age Hut in Israel Yields World's Oldest Evidence of Bedding', *Proc. Natl. Acad. Sci. USA* 101(2004): 6821–6.

8. This return to Ice Age conditions is generally known as the Younger Dryas; W. J. Burroughs, *Climate Change in Prehistory: The End of the Reign of Chaos* (Cambridge: Cambridge University Press, 2005).

9. The post-glacial colonization of Europe from the south-west has been well documented by tracking the routes followed by genetic markers. A. Torroni, et al., 'MtDNA Analysis Reveals a Major Late Palaeolithic Population Expansion from Southwestern to Northeastern Europe', *Am. J. Hum. Genet.* 62(1998): 1137–52; A. Torroni et al., 'A Signal, from Human mtDNA, of Postglacial Recolonization in Europe', *Am. J. Hum. Genet.* 69(2001): 844–52.

10. Mithen, *After the Ice*, provides a comprehensive account of the period of post-glacial settlement of humans across the world.

11. A. Currrey, 'Seeking the Roots of Ritual', *Science* 319(2008): 278–80.

12. The earliest known domesticated wheat has been found at the Turkish site of Nevali Çori, not far to the north-west of Göbekli Tepe and dated to 10.5

thousand years ago. M. Balter, 'Seeking Agriculture's Ancient Roots', *Science* 316(2007): 1830–5.

13. K. Tanno and G. Willcox, 'How Fast Was Wild Wheat Domesticated?', *Science* 311(2006): 1886.

14. Harvested plants included acorns, pistachios, olives, and large quantities of wild wheat and barley but no cultivation was involved; Balter, 'Seeking Agriculture's Ancient Roots'.

15. M. A. Zeder, 'Central Questions in the Domestication of Plants and Animals', *Evol. Anthropol.* 15(2006): 105–17.

16. J. Diamond, 'Evolution, Consequences and Future of Plant and Animal Domestication', *Nature* 418(2002): 700–7.

Epilogue: Children of Chance

1. T. H. Clutton-Brock and P. Harvey, 'Primates, Brains, and Ecology', *J. Zool.* 190(1980): 309–23; P. H. Harvey, T. H. Clutton-Brock, and G. M. Mace, 'Brain Size and Ecology in Small Mammals and Primates', *Proc. Natl. Acad. Sci. USA* 77(1980): 4387–9.

2. A. A. S. Weir, J. Chappell, and A. Kacelnik, 'Shaping of Hooks in New Caledonian Crows', *Science* 297(2002): 981; F. B. M. de Waal and P. L. Tyack (eds), *Animal Social Complexity: Intelligence, Culture, and Individualized Societies* (Cambridge, MA: Harvard University Press, 2003); N. J. Emery et al., 'The Mentality of Crows: Convergent Evolution of Intelligence in Corvids and Apes', *Science* 306(2004): 1903; N. J. Emery et al., 'Cognitive Adaptations of Social Bonding in Birds', *Phil. Trans. Roy. Soc. B* 362(2007): 489–505; K. E. Holekamp, S. T. Sakai, and B. L. Lundrigan, 'Social Intelligence in the Spotted Hyena (*Crocuta crocuta*)', *Phil. Trans. Roy. Soc. B* 362(2007): 523–38; J. A. Mather, 'Cephalopod Consciousness: Behavioural Evidence', *Consc. Cogn.* 17(2008): 37–48.

3. L. C. Aiello and R. I. M. Dunbar, 'Neocortex Size, Group Size, and the Evolution of Language', *Curr. Anthropol.* 34(1993): 184–93; R. I. M. Dunbar, 'THE SOCIAL BRAIN: Mind, Language, and Society in Evolutionary Perspective', *Ann. Rev. Anthropol.* 32(2003): 163–81.

4. C. P. van Schaik and R. O. Deaner, 'Life History and Cognitive Evolution in Primates', in de Waal and Tyack (eds), *Animal Social Complexity*, 5–25.

5. Advantages in obtaining patchily distributed food by living in groups include improved food location, improved chance of catching prey, ability to catch larger prey, and competing better for food against other species. Advantages in avoiding predation include avoiding detection, detecting the predator, deterring the predator, confusing the predator, diluting the predator's effects, and avoiding becoming a victim; B. C. R. Bertram, 'Living in Groups: Predators and Prey', in J. R. Krebs and N. B. Davies (eds), *Behavioural Ecology: An Evolutionary Approach* (Oxford: Blackwell, 1978), 64–96.

6. L. C. Aiello and P. W. Wheeler, 'The Expensive-Tissue Hypothesis', *Curr. Anthropol.* 36(1995): 199–221.

7. C. B. Stanford and H. T. Bunn (eds), *Meat-Eating and Human Evolution* (New York: Oxford University Press, 2001).

8. Climate-driven ecological change edging *Homo erectus* and *H. heidelbergensis* away from woodland and onto the wooded savannahs and steppes; meat, fat, and marrow consumption making large brains possible.

9. M. Ponce de León et al., 'Neanderthal Brain Size at Birth Provides Insights into the Evolution of Human Life History', *Proc. Natl. Acad. Sci. USA* 105(2008): 13764–8.

10. Later humans were smaller and had correspondingly smaller brains than Neanderthals or early humans; C. B. Ruff, E. Trinkaus, and T. W. Holliday, 'Body Mass and Encephalization in Pleistocene *Homo*', *Nature* 387(1997): 173–6.

11. A. B. Migliano, L. Vinicius, and M. M. Lahr, 'Life History Trade-offs Explain the Evolution of Human Pygmies', *Proc. Natl. Acad. Sci. USA* 104(2007): 20216–19.

12. The cerebral hemispheres within the brain play a central role in perceptual awareness, memory, attention, thought, consciousness, and language while the cerebellum integrates sensory perception, motor control, and coordination. It seems that the re-organization of the brain in, at least some, Late Pleistocene and Holocene humans involved a development of the cerebellum at the expense of the cerebral hemispheres; A. H. Weaver, 'Reciprocal Evolution of the Cerebellum and Neocortex in Fossil Humans', *Proc. Natl. Acad. Sci. USA* 102(2005): 3576–80.

13. J. Kien, 'The Need for Data Reduction May Have Paved the Way for the Evolution of Language Ability in Hominids', *J. Hum. Evol.* 20(1991): 157–65.

14. Changes in cerebellum development need not have been exclusively genetic in origin. For example, observed differences in cerebellar volume between musicians and non-musicians may be the result of adaptation to the rigours of musical training, possibly at a particular stage of brain development in the individual; S. Hutchinson et al., 'Cerebellar Volume of Musicians', *Cereb. Cortex* 13(2003): 943–9.

15. J. Diamond, *Guns, Germs and Steel. A Short History of Everybody for the Last 13,000 Years* (London: Jonathan Cape, 1997).

16. And presumably also demographically although it is next to impossible to estimate population numbers from a scattering of fossils.

17. P. S. Martin and R. G. Klein, *Quaternary Extinctions: A Prehistoric Revolution* (Tucson: University of Arizona Press, 1984).

18. M. C. Stiner et al., 'Paleolithic Population Growth Pulses Evidenced by Small Animal Exploitation', *Science* 283(1999): 190–4; M. C. Stiner, N. D. Munro, and T. A. Surovell, 'The Tortoise and the Hare: Small-Game Use, the Broad Spectrum Revolution, and Paleolithic Demography', *Curr. Anthropol.* 41(2000): 39–74.

ENDNOTES

19. C. Finlayson, *Neanderthals and Modern Humans: An Ecological and Evolutionary Perspective* (Cambridge: Cambridge University Press, 2004).

20. Mather, 'Cephalopod Consciousness'.

21. J. M. Plotnik, F. B. de Waal, and D. Reiss, 'Self-Recognition in an Asian Elephant', *Proc. Natl. Acad. Sci. USA* 103(2006): 17053–7.

22. D. Reiss and L. Marino, 'Mirror Self-Recognition in the Bottlenose Dolphin: A Case of Cognitive Convergence', *Proc. Natl. Acad. Sci. USA* 98(2001): 5937–42.

23. G. G. Gallup, Jr., 'Chimpanzees: Self-Recognition', *Science* 167(1970): 86–7; S. D. Suarez and G. G. Gallup, Jr., 'Self-Recognition in Chimpanzees and Orangutans, But Not Gorillas', *J. Hum. Evol.* 10(1981): 175–88; D. J. Povinelli et al., 'Self-Recognition in Chimpanzees (*Pan troglodytes*): Distribution, Ontogeny, and Patterns of Emergence', *J. Comp. Psychol.* 107(1993): 347–72; V. Walraven, L. van Elsacker, and R. Verheyen, 'Reactions of a Group of Pygmy Chimpanzees (*Pan paniscus*) to Their Mirror-images: Evidence of Self-Recognition', *Primates* 36(1995): 145–50; D. J. Povinelli et al., 'Chimpanzees Recognize Themselves in Mirrors', *Anim. Behav.* 53(1997): 1083–8.

24. H. M. Leach, 'Human Domestication Reconsidered', *Curr. Anthropol.* 44(2003): 349–68.

25. J. Tooby and L. Cosmides, 'The Past Explains the Present. Emotional Adaptations and the Structure of Ancestral Environments', *Ethol. Sociobiol.* 11(1990): 375–424; S. B. Eaton, S. B. Eaton III, and M. J. Konner, 'Paleolithic Nutrition Revisited: A Twelve-Year Retrospective on Its Nature and Implications', *Eur. J. Clinic. Nutr.* 51(1997): 207–16; P. Shepard, *Coming Home to the Pleistocene*, (Washington: Island Press, 1998); C. M. Pond, *The Fats of Life* (Cambridge: Cambridge University Press, 1998); F. W. Booth, M. V. Chakravarty, and E. E. Spangenburg, 'Exercise and Gene Expression: Physiological Regulation of the Human Genome through Physical Activity', *J. Physiol.* 543(2002): 399–411; L. Cordain et al., 'Origins and Evolution of the Western Diet: Health Implications for the 21st Century', *Am. J. Clin. Nutr.* 81(2005): 341–54; P. Gluckman and M. Hanson, *MisMatch. Why Our World No Longer Fits Our Bodies* (Oxford: Oxford University Press, 2006).

Index

INDEX